1,000,000 Books

are available to read at

www.ForgottenBooks.com

Read online
Download PDF
Purchase in print

ISBN 978-1-331-92767-9
PIBN 10255507

This book is a reproduction of an important historical work. Forgotten Books uses state-of-the-art technology to digitally reconstruct the work, preserving the original format whilst repairing imperfections present in the aged copy. In rare cases, an imperfection in the original, such as a blemish or missing page, may be replicated in our edition. We do, however, repair the vast majority of imperfections successfully; any imperfections that remain are intentionally left to preserve the state of such historical works.

Forgotten Books is a registered trademark of FB &c Ltd.
Copyright © 2018 FB &c Ltd.
FB &c Ltd, Dalton House, 60 Windsor Avenue, London, SW19 2RR.
Company number 08720141. Registered in England and Wales.

For support please visit www.forgottenbooks.com

1 MONTH OF FREE READING

at
www.ForgottenBooks.com

By purchasing this book you are eligible for one month membership to ForgottenBooks.com, giving you unlimited access to our entire collection of over 1,000,000 titles via our web site and mobile apps.

To claim your free month visit:
www.forgottenbooks.com/free255507

* Offer is valid for 45 days from date of purchase. Terms and conditions apply.

English
Français
Deutsche
Italiano
Español
Português

www.forgottenbooks.com

Mythology Photography **Fiction** Fishing Christianity **Art** Cooking Essays Buddhism Freemasonry Medicine **Biology** Music **Ancient Egypt** Evolution Carpentry Physics Dance Geology **Mathematics** Fitness Shakespeare **Folklore** Yoga Marketing **Confidence** Immortality Biographies Poetry **Psychology** Witchcraft Electronics Chemistry History **Law** Accounting **Philosophy** Anthropology Alchemy Drama Quantum Mechanics Atheism Sexual Health **Ancient History Entrepreneurship** Languages Sport Paleontology Needlework Islam **Metaphysics** Investment Archaeology Parenting Statistics Criminology **Motivational**

| 32d Congress, | SENATE. | Executive, |
| 2d Session. | | No. 54. |

EXPLORATION

OF THE

RED RIVER OF LOUISIANA,

IN THE YEAR 1852;

BY

RANDOLPH B. MARCY,
CAPTAIN FIFTH INFANTRY U. S. ARMY;

ASSISTED BY

GEORGE B. McCLELLAN,
BREVET CAPTAIN U. S. ENGINEERS.

WITH REPORTS ON THE NATURAL HISTORY OF THE COUNTRY,
AND NUMEROUS ILLUSTRATIONS.

WASHINGTON:
ROBERT ARMSTRONG, PUBLIC PRINTER.
1853.

REPORT

OF

THE SECRETARY OF WAR,

COMMUNICATING,

In compliance with a resolution of the Senate, Captain Marcy's Report of his exploration of the Red river.

FEBRUARY 4, 1853.—Ordered to be printed.
MARCH 10.—Ordered that 2,000 additional copies be printed, 200 of which for Captain Marcy.

WAR DEPARTMENT,
Washington, November 8, 1853.

SIR: In compliance with a resolution of the Senate of the 4th of February, 1853, I have the honor to transmit herewith, for the use of the Senate, a copy of the report of Captain R. B. Marcy of his exploration of the waters of Red river.

Very respectfully, your obedient servant,

JEFFN. DAVIS,
Secretary of War.

ASBURY DICKINS, Esq.,
Secretary of the Senate.

INTRODUCTION.

In submitting the following report of a reconnoissance of the country bordering upon upper Red river, it is proper to state that previous to our departure upon the expedition, we were unable to procure all the instruments adapted to the performance of such services as were required of us. We succeeded in obtaining a sextant, a mountain barometer, an aneroid barometer, an odometer, a prismatic compass, and two Fahrenheit thermometers; but could not procure a chronometer, and, in consequence, were under the necessity of making our observations with a pocket lever watch.

The latitudes given are the results of from twelve to fifteen observations of Polaris for the determination of each position. The longitudes were determined by a series of observations upon lunar distances, and are believed to be as accurate as the imperfect character of our instruments would admit.

The positions thus deduced have been corrected by frequent and careful observations of courses and distances with the compass and odometer, a record of which will be found in the appendix.

The astronomical observations were made by Captain George B. McClellan, of the engineer corps, who, in addition to the duties properly pertaining to his department, performed those of quartermaster and commissary to the command. An interesting collection of reptiles and other specimens, in alcohol, was also made under his superintendence, and put into the hands of Professors Baird and Girard, of the Smithsonian Institution, whose reports will be found in the appendix. For these and many other important services, as well as for his prompt and efficient co-operation in whatever was necessary for the successful accomplishment of the design of the expedition, I take this opportunity of tendering my warmest acknowledgment.

Doctor George G. Shumard, of Fort Smith, Arkansas, who faithfully discharged the duties of surgeon to the command, also made important contributions to the department of natural science, by collections of specimens of the rocks, minerals, soils, fossils, shells, and plants, of the different localities which we traversed; and of these, the plants were placed in the hands of Dr. John Torrey, of New York, the eminent botanist so well known to the army by his able reports on the collections of Fremont, Emory, and others.

The shells were intrusted to Professor C. B. Adams, of Amherst. His report, as presented, possesses a melancholy interest, as being almost the last scientific effort of this distinguished conchologist, whose loss science has so recently been called upon to deplore.

The specimens of rocks and minerals have been examined by President Hitchcock, of Amherst College, with important results, while copious remarks on the general geology of the country have been supplied by Dr. Shumard, who has also furnished some notes on the conchology of the route.

The minerals and soils have been analyzed by Professor C. U. Shepard, who detected among them a new species. Finally, in the hands of Dr. Benjamin F. Shumard, the fossils have yielded several novelties to science. All these reports upon the natural history of the expedition will be found detailed at length in the appendix.

The barometrical observations which are given were taken from both forms of the instruments, and exhibited a remarkable agreement until the 8th of June, when we had the misfortune to break the mountain barometer, and were obliged subsequently to depend solely upon the aneroid. This I believe to be very reliable, as it has been tested since our return by a careful comparison with several other instruments in possession of Benjamin Pike & Son, New York, and found to be in perfect order.

In order to obtain as intimate a knowledge as possible of the country over which we passed, I was necessarily absent from the train a great portion of the time while it was in motion; and during such periods the command devolved upon

Lieutenant Updegraff, which, with the constant guard I deemed it necessary to keep over our animals in a country where the Indians manifested a disposition by no means friendly towards us, made his varied duties laborious, and it gives me pleasure to bear testimony to the efficient manner in which he performed them.

R. B. MARCY,
Captain 5th Infantry.

TABLE OF CONTENTS.

CHAPTER I.

Order from headquarters of the army—Failure of former expeditions in reaching the source of Red river—Causes of failure—Departure from Washington—Arrival at Fort Belknap—The Little Witchita—Big Witchita—Departure from Cache creek—Copper ore—Indian signs - - - - - - - - - - - - - - - - - - - Page 1

CHAPTER II.

Witchita mountains—Panther killed—Buffalo traces—Singular and unaccountable rise of water—Buffalo signs—Horse captured—Rains—Arrival at Otter creek—Barometer broken—Character of Witchita mountains—Buffalo killed—High water - - - - - - - Page 10

CHAPTER III.

Witchitas—Discouraging account of the country in advance—Pass 100° of longitude—Leave Otter creek—Berries—Elk creek—Pass Witchita mountains—Gypsum bluffs—Buffaloes seen—Suydam creek—Comanche signs - - - - - - - - - - - - - - - Page 17

CHAPTER IV.

Buffalo chase—Sweet Water creek—Comanche camps—Prevailing winds—Indians seen—Method of encamping—Wonderful powers of the Delawares—Beaver dams—Kioway creek - - - - - - Page 27

CHAPTER V.

Reach the source of the north branch of Red river—Bottle buried—Arrived upon the Canadian—Departure for Middle Fork—Indian battleground—Prairie-dog towns—Source of Middle Fork—South Fork—Prairie dogs - - - - - - - - - - - - - - - - Page 37

CHAPTER VI.

Arrive at the main South Fork—Panther killed—Bitter water—Intense thirst—Head spring—Bears abundant—Departure down the river - - - - - - - - - - - - - - - - - Page 49

CHAPTER VII.

Antelope and deer—Witchita mountains in sight—Reach Buffalo creek—Valley of Otter creek—Salubrity of climate—Deer-bleat—Horseflies—Scurvy—Witchita mountains—Pass through the mountains—Buffalo seen - - - - - - - - - - - - - - Page 62

CHAPTER VIII.

Old Indian villages—Beautiful scenery—Trap formation—Lost mule—Beaver creek—Prairie guides—Rush creek—Witchita and Waco villages—Mexican prisoners—Talk with the Indians—Cross Timbers—Kickapoos—Strike wagon track—Arrival at Fort Arbuckle. Page 72

CHAPTER IX.

Prominent features of the Red river—Chain of lakes—Cross Timbers—Arable lands—Establishment of a military post upon Red river recommended—Route of Comanches and Kioways in passing to Mexico—Wagon route from Fort Belknap to Santa Fe—Navigation of Red river—Erroneous opinions in regard to Red river—Extensive gypsum range—El Llano Estacado - - - - - - - - - - - Page 83

CHAPTER X.

Indians of the country—Habits of Comanches and Kioways—Similarity between them and the Arabs and Tartars—Predatory excursions into Mexico—War implements—Incredulity regarding the customs of the whites—Method of saluting strangers—Degraded condition of the women—Aversion to ardent spirits—Prairie Indians contrasted with the Indians of Eastern States—Buffaloes—Probable condition of Indians on the extermination of the buffaloes—Pernicious influence of traders—Superstitions of the natives - - - - - - - - - - - - Page 93

CHAPTER XI.

Pacific railway—Impracticability of crossing the "Llano Estacado"—Route from Fort Smith to Santa Fe—Return route from Doña Ana—Its connexions with the Mississippi and Pacific - - - - Page 109

APPENDIX A.

Meteorological Observations - - - - - - - - - Page 119

APPENDIX B.

Tables of Courses and Distances - - - - - - Page 139

APPENDIX C.

Mineralogy:
    Report on the minerals collected. By Prof. C. U. Shepard - - - - - - - - - - - - - - - - Page 155

APPENDIX D.

Geology:
    Notes upon the specimens of rocks and minerals collected. By President Edward Hitchcock - - - - - - Page 163
    Remarks upon the general geology of the country traversed. By George G. Shumard, M. D. - - - - - - Page 179

APPENDIX E.

Palæontology:
    Description of the species of carboniferous and cretaceous fossils collected. By B. F. Shumard, M. D. - - - - Page 199

APPENDIX F.

Zoology:
    Mammals. By R. B. Marcy, Captain U. S. A. - - Page 215
    Reptiles. By S. F. Baird and C. Girard - - - - Page 217
    Fishes. By S. F. Baird and C. Girard - - - - Page 245
    Shells. By C. B. Adams and G. G. Shumard, M. D. Page 253
    Orthopterous insects. By C. Girard - - - - - - Page 257
    Arachnidians. By C. Girard - - - - - - - Page 262
    Myriapods. By C. Girard - - - - - - - - Page 272

APPENDIX G.

Botany:
    Description of the plants collected during the expedition. By Dr. John Torrey - - - - - - - - - - Page 279

APPENDIX H.

ETHNOLOGY:
  Vocabulary of the Comanches and Witchitas. By Captain R. B. Marcy; with some general remarks by Prof. W. W. Turner - - - - - - - - - - - - - - - Page 307

ALPHABETICAL INDEX - - - - - - - - - - Page 313

ILLUSTRATIONS.

LANDSCAPES.

Plate.		Page.
I.	Granite boulders	22
II.	Mount Webster	21
III.	Encampment of 6th June	25
IV.	Gypsum Bluffs on north branch of Red river	23
V.	Views of Bluffs on Canadian river	39
VI.	View near Gypsum Bluffs on Red river	24
VII.	Border of El Llano Estacado	50
VIII.	View near head of Red river	55
IX.	View near head of Red river	55
X.	Head of main branch of Red river	56
XI.	Trap mountain on Cache creek	73
XII.	Witchita village on Rush creek	78

GEOLOGICAL SECTIONS.*

I.	Section showing the order and succession of the strata from Washington county, Arkansas, to Fort Belknap, Texas	179
III.	Section on Cache creek, near its junction with Red river	182
IV.	Section of strata on north branch of Red river, taken June 2	187
V.	Out-crop of finely laminated ferruginous sandstone near north branch of Red river	189
VI.	Section of the borders of the Llano Estacado, taken June 16	190
VII.	Section of strata near middle branch of Red river, taken June 21	190
VIII.	Section of bluffs between middle branch of Red river and Dog-town river, taken June 24	191

*All the geological sections are by Dr. George G. Shumard, surgeon to the expedition.

Plate.		Page.
IX.	Section of strata near the head of Red river, taken June 28	191
X.	Section of cliffs on Cache creek - - - - - - - - -	194

PALÆONTOLOGY.*

I.	Fig. 1 a. Productus cora, D'Orb. - - - - - - - -	202
	Fig. 2. Productus costatus, Sow. - - - - - - - -	202
	Fig. 3. Spirifer, *indet.* - - - - - - - - - - -	203
	Fig. 4 a. b. Terebratula marcyi, Shum. - - - - - -	203
	Fig. 5. Productus punctatus, Martin - - - - - - -	201
	Fig. 6. Archimedipora archimedes, Les. - - - - - -	201
	Fig. 7. Agassizocrinus dactyliformis, Troost - - - -	199
II.	Fig. 1. Productus punctatus, Martin - - - - - - -	201
	Fig. 2 a. b. Pecten quadricostatus, Sow. - - - - -	204
	Fig. 3 a. b. Terebratula choctawensis, Shum. - - - -	207
	Fig. 4 a. b. c. Hemiaster elegans, Shum. - - - - -	210
III.	Fig. 1. Ammonites acuto-carinatus, Shum. - - - - -	209
	Fig. 2. Holaster simplex, Shum. - - - - - - - -	210
	Fig. 3. Astarte washitensis, Shum. - - - - - - -	206
	Fig. 4. Ammonites, *indet.* - - - - - - - - - -	210
	Fig. 5. Exogyra texana, Roem. - - - - - - - - -	205
	Fig. 6. Pecten quadricostatus, Sow. - - - - - - -	204
IV.	Fig. 1. Trigonia crenulata, Lam. - - - - - - - -	206
	Fig. 2. Cardium multistriatum, Shum. - - - - - -	207
	Fig. 3. Eulima subfusiformis, Shum. - - - - - - -	208
	Fig. 4. Globiconcha elevata, Shum. - - - - - - -	208
	Fig. 5. Ammonites marcianus, Shum. - - - - - -	209
	Fig. 6. Pileopsis; not mentioned in the text - - - -	
	Fig. 7. Holectypus planatus, Roem. - - - - - - -	211
	Fig. 8. Terebratula subtilita, Hall - - - - - - -	202
V.	Fig. 1 a. b. Exogyra texana, Roem. - - - - - - -	205
	Fig. 2. Ostrea subovata, Shum. - - - - - - - -	205

---

* Owing to the impossibility of communicating with Dr. Shumard during the printing and engraving of the present report, I have been unable to fill up the gaps in the above list of figures made up from the references in the article on palæontology. This want of supervision on the part of the author will also explain the existence of sundry discrepancies between text and plates.

R. B. M.

ILLUSTRATIONS.

Plate.		Page.
V. Fig. 3. Globiconcha (Tylostoma) tumida, Shum.		208
VI. Fig. 1. Panopæa texana, Shum.		207
Fig. 2. Inoceramus confertim-annulatus, Roem.		206
Fig. 3.		
Fig. 4. Natica, *indet.* (cast); not mentioned in the text		
Fig. 5. Gryphæa pitcheri, Morton		205

ZOOLOGY.

I. Crotalus confluentus, Say		217
II. Eutænia proxima, B. & G.		220
III. Eutænia marciana, B. & G.		221
IV. Heterodon nasicus, B. & G.		222
V. Pituophis mcclellanii, B. & G.		225
VI. Scotophis lætus, B. & G.		227
VII. Ophibolus sayi, B. & G.		228
VIII. Ophibolus gentilis, B. & G.		229
IX. Leptophis majalis, B. & G.		232
X. Figs. 1–4. Cnemidophorus gularis, B. & G.		239
Figs. 5–12. Sceloporus consobrinus, B. & G.		237
XI. Bufo cognatus, Say		242
XII. Pomotis (Bryttus) longulus, B. & G.		245
XIII. Pomotis breviceps, B. & G.		246
XIV. Figs. 1–4. Leuciscus vigilax, B. & G.		248
Figs. 5–8. Leuciscus bubalinus, B. & G.		249
Figs. 9–12. Leuciscus lutrensis, B. & G.		251
XV. Figs. 1–4. Brachypeplus magnus, G.		260
Figs. 5–8. Anabrus haldemanii, G.		259
Figs. 9–13. Daihinia brevipes, Hald.		257
XVI. Figs. 1–3. Mygale hentzii, G.		262
Figs. 4–5. Lycosa pilosa, G.		263
XVII. Figs. 1–4. Thelyphonus excubitor, G.		265
Figs. 5–7. Scorpio (Telegonus) boreus, G.		267

Plate.	Page.
XVIII. Scolopendra heros, G.	272

BOTANY.*

Plate	Page
I. Anemone caroliniana, Walt.	280
II. Dithyræa wislizenii, Engelm.	280
III. Geranium Fremontii, Torr.	
IV. Hoffmanseggia Jamesii, Torr. & Gr.	284
V. Sanguisorba annua, Nutt.	285
VI. Eryngium diffusum, Torr.	286
VII. Eurytænia Texana, Torr. & Gr.	287
VIII. Liatris acidota, Engelm. & Gray	287
IX. Aphanostephus ramosissimus, DC.	289
X. Xanthisma Texana	
XI. Engelmannia pinnatifida, Torr. & Gr.	289
XII. Artemisia filifolia, Torr.	287
XIII. Erythræa Beyrichi, Torr. & Gr.	291
XIV. Heliotropium tenellum	
XV. Euploca convolvulacea, Nutt.	294
XVI. Pentstemon ambiguus, Torr.	292
XVII. Lippia cuneifolia, Torr.	293
XVIII. Abronia cycloptera	
XIX. Poa interrupta	301
XX. Uniola stricta, Torr.	301

* For explanations of the figures on each plate, see p. 303.

MAPS.

1. Map of the country between the frontiers of Arkansas and New Mexico; embracing the section explored in 1849, –'50, –'51, and –'52, by Captain R. B. Marcy, 5th U. S. infantry, under orders from the War Department. Also, a continuation of the emigrant road from Fort Smith and Fulton down the valley of the Gila.

2. Map of the country embraced within the basin of Upper Red river, explored in 1852 by Captain R. B. Marcy, fifth infantry, assisted by Brevet Captain George B. McClellan, U. S. engineers.

CHAPTER I.

ORDER FROM HEADQUARTERS OF THE ARMY—FAILURE OF FORMER EXPEDITIONS IN REACHING THE SOURCES OF RED RIVER—CAUSES OF FAILURE—DEPARTURE FROM WASHINGTON—ARRIVAL AT FORT BELKNAP—THE LITTLE WITCHITA—BIG WITCHITA—DEPARTURE FROM CACHE CREEK—COPPER ORE—INDIAN SIGNS.

NEW YORK, *December* 5, 1852.

Col. S. COOPER, *Adjutant General U. S. Army:*

SIR: I have the honor herewith to submit a report of an exploration of the country embraced within the basin of Upper Red river, made in obedience to the following orders:

[SPECIAL ORDERS No. 33.]   ADJUTANT GENERAL'S OFFICE,
Washington, March 5, 1852.

Captain R. B. Marcy, 5th Infantry, with his company as an escort, will proceed, without unnecessary delay, to make an examination of the Red river, and the country bordering upon it, from the mouth of Cache creek to its sources, according to the special instructions with which he will be furnished. On completing the exploration, Captain Marcy will proceed to Washington to prepare his report.

Brevet Captain G. B. McClellan, Corps of Engineers, is assigned to duty with this expedition. Upon the completion of the field service he will report to Brevet Major General Smith, the commander of the 8th department.

The necessary supplies of subsistence and quartermasters' stores will be furnished from the most convenient depots in the 7th or 8th military department.

By command of Major General Scott:

R. JONES,
Adjutant General.

Before proceeding to give a detailed account of the expedition, it may be proper to remark, that during the greater portion of the three years previous to the past summer I had been occupied in exploring the district of country lying upon the Canadian river of the Arkansas, and upon the headwaters of the Trinity, Brazos, and Colorado rivers of Texas.

During this time my attention was frequently called to the remarkable fact that a portion of one of the largest and most important rivers in the United States, lying directly within the limits of the district I had been examining, remained up to that late period wholly unexplored and unknown, no white man having ever ascended the stream to its sources. The only information we had upon the subject was derived from Indians and semi-civilized Indian traders, and was of course very unreliable, indefinite, and unsatisfactory; in a word, the country embraced within the basin of Upper Red river had always been to us a "terra incognita." Several enterprising and experienced travellers had at different periods attempted the examination of this river, but, as yet, none had succeeded in reaching its sources.

At a very early period, officers were sent out by the French government to explore Red river, but their examinations appear to have extended no further than the country occupied by the Natchitoches and Caddoes in the vicinity of the present town of Natchitoches, Louisiana. Subsequent examinations had extended our acquaintance with its upper tributaries, but we were still utterly in the dark in regard to the true geographical position of its sources.

Three years after the cession to the United States, by the First Consul of the French republic, of that vast territory then known as Louisiana, a small party, called the "Exploring expedition of Red river," consisting of Capt. Sparks, Mr. Freeman, Lieut. Humphry, and Dr. Custis, with seventeen private soldiers, two non-commissioned officers, and a black servant, embarked from Saint Catherine's landing near Natchez, Mississippi, with instructions to ascend Red river to its sources. They descended the Mississippi, and on the 3d of May, 1806, entered Red river, expecting to be able to ascend in their boats to the country of the Pawnee (Pique) Indians. Here it was their intention to leave their boats, and, after packing provisions on horses which they were to purchase from the Pawnees, to proceed (as expressed in their orders) *to the top of the mountains*, the distance being, as they conjectured, about three hundred miles.

It is evident from the foregoing that Red river was supposed to issue from a mountainous country, and the preparations for this expedition were made accordingly. This party encountered many difficulties and obstructions in the navigation of the river among the numerous bayous in the vicinity of the great raft, but finally overcame them all, and found themselves upon the river above this formidable obstacle. They were, however, soon met by a large force of Spanish troops, the commander of which ordered them to proceed no further; and as their numbers were too small for a thought of resistance, they were forced to turn back and abandon the enterprise.

Another expedition was fitted out in 1806 by our government and placed under the command of that enterprising young traveller, Lieut.

Pike, who was ordered to ascend the Arkansas river to its sources, thence to strike across the country to the head of Red river, and descend that stream to Natchitoches. After encountering many privations and intense sufferings in the deep snows of the lofty mountains about the headwaters of the Arkansas, Lieut. Pike arrived finally upon a stream running to the east, which he took to be Red river, but which subsequently proved to be the Rio Grande. Here he was taken by the governor of New Mexico and sent home by way of Chihuahua and San Antonio, thus putting a stop to his explorations.

General Wilkinson, under whose orders Lieut. Pike was serving at the time, states, in a letter to him after his return, as follows: "The principal object of your expedition up the Arkansas was to discover the true position of the sources of Red river. This was not accomplished." Lieut. Pike, however, from the most accurate information he could obtain, gives the geographical position of the sources of Red river as in latitude 33° N. and longitude 104° W. Again, in 1819–'20, Col. Long, of the U. S. Topographical Engineers, on his return from an exploration of the Missouri river and the country lying between that stream and the head of the Arkansas, undertook to descend the Red river from its sources. The Colonel, in speaking of this in his interesting report, says: "We arrived at a creek having a westerly course, which we took to be a tributary of Red river. Having travelled down its valley about two hundred miles, we fell in with a party of Indians, of the nation of "Kaskias," or "Bad Hearts," who give us to understand that the stream along which we were travelling was Red river. We accordingly continued our march down the river several hundred miles further, when, to our no small disappointment, we discovered it was the Canadian of the Arkansas, instead of Red river, that we had been exploring.

"Our horses being nearly worn out with the fatigue of our long journey, which they had to perform bare-footed, and the season being too far advanced to admit of our retracing our steps and going back again in quest of the source of Red river with the possibility of exploring it before the commencement of winter, it was deemed advisable to give over the enterprise for the present and make our way to the settlements on the Arkansas. We were led to the commission of this mistake in consequence of our not having been able to procure a good guide acquainted with that part of the country. Our only dependence in this respect was upon Pike's map, which assigns to the headwaters of Red river the apparent locality of those of the Canadian."

Doctor James, who accompanied Colonel Long, in his journal of the expedition, says: "Several persons have recently arrived at St. Louis, in

Missouri, from Sante Fé, and among others the brother of Captain Shreeves, who gives information of a large and frequented road, which runs nearly due east from that place, and strikes one of the branches of the Canadian, that, at a considerable distance south of this point, in the high plain, is the principal source of Red river.

"His account confirms an opinion we had previously formed, namely: that the branch of the Canadian explored by Major Long's party in August, 1820, has its sources near those of some stream which descends towards the west into the Rio del Norte, and consequently that some other region must contain the head of Red river." He continues:

"From a careful comparison of all the information we have been able to collect, we are satisfied that the stream on which we encamped on the 31st of August is the Rio Raijo of Humboldt, long mistaken for the sources of Red river of Natchitoches. In a region of red clay and sand, where all the streams become nearly the color of arterial blood, it is not surprising that several rivers should have received the same name; nor is it surprising that so accurate a topographer as the Baron Humboldt, having learned that a Red river rises forty or fifty miles east of Santa Fé and runs to the east, should conjecture it might be the source of Red river of Natchitoches.

"This conjecture (for it is no more) we believe to have been adopted by our geographers, who have with much confidence made their delineations and their accounts to correspond with it."

Hence it will be seen that up to this time there is no record of any traveller having reached the sources of Red river, and that the country upon the headwaters of that stream has heretofore been unexplored. The Mexicans and Indians on the borders of Mexico are in the habit of calling any river, the waters of which have a red appearance, "Rio Colorado," or Red river, and they have applied this name to the Canadian in common with several others; and as many of the prairie Indians often visit the Mexicans, and some even speak the Spanish language, it is a natural consequence that they should adopt the same nomenclature for rivers, places, &c. Thus, if a traveller in New Mexico were to inquire for the head of Red river, he would most undoubtedly be directed to the Canadian, and the same would also be the case in the adjacent Indian country. These facts will account for the mistake into which Baron Humboldt was led, and it will also account for the error into which Colonel Long and Lieut. Pike have fallen in regard to the sources of the stream which we call Red river.

Dr. Gregg, in his "Commerce of the Prairies," tells us that on his way down the south bank of the Canadian his Comanche guide, Manuel, (who, by-the-by, travelled six hundred miles with me upon the plains, and whom I always found reliable,) pointed out to him breaks or bluffs upon a stream to the south of the Canadian, near what we ascertained to be the true position of the head of the north branch of Red river, and where it approaches within twenty-five miles of the Canadian. These bluffs he said were upon the "Rio Negro," which the Doctor supposed to be the Washita river; but after having examined that section of country I am satisfied that the north branch of Red river must have been alluded to by the guide, as the Washita rises further to the east. It therefore seems probable that "Rio Negro" is the name which the Mexicans have applied to Red river of Louisiana.

Immediately on the receipt of the foregoing order I repaired to Fort Smith, Arkansas, where the Quartermaster General had directed that transportation should be furnished me, but on arriving there I learned that nearly all the means of transportation had a short time before been transferred to the depot at Preston, Texas. Captain Montgomery, the quartermaster at Fort Smith, manifested every disposition to facilitate my movements, and supplied me with ten most excellent horses, with which I proceeded on to Preston. At this point I made a requisition upon the quartermaster for a sufficient number of teams to transport supplies of subsistence, and baggage for my command, for five months. These were promptly furnished by Bvt. Major George Wood, to whom I am under many obligations for his active and zealous co-operation in supplying me with such articles as were necessary for the expedition. With but few resources at his command, with animals that had been worked down, and, in consequence of the scarcity of grain, very poor, and with parts of old wagons much worn, he succeeded in a very few days in fitting me out with twelve ox teams that performed very good service.

As my company was at Fort Belknap, upon the Brazos river, one hundred and sixty miles from Preston, and as the route by way of Fort Arbuckle to the mouth of Cache creek (the initial point of my reconnoissance upon Red river) is much the shortest, I determined to leave my supply train under the charge of a wagonmaster to bring forward over this route, and to proceed myself to Fort Belknap and march my company over the other trail, uniting with the train at the mouth of Cache creek.

I accordingly reached Fort Belknap on the 30th of April, and on the 2d of May left with my company, marching over the Fort Arbuckle road as far as where it intersects Red river. As our road led us along near the valley of the Little Witchita, I took occasion to examine it more particularly than I had ever done before, and found it a much more desirable section of country than I had imagined.

The soil in the valley is very productive; the timber, consisting of overcup, white-oak, elm, hackberry, and wild china, is large and abundant, and the adjoining prairie is covered with a heavy growth of the very best grass. The stream at fifteen miles above its confluence with Red river is twenty feet wide and ten inches deep, with a rapid current, the water clear and sweet.

From the point where I first struck it, good farms could be made along the whole course of the creek to its mouth. The country adjoining is high, rolling prairie, interspersed here and there with groves of post-oak, and presents to the eye a most pleasing appearance.

From the Little Witchita we ascended Red river along the south bank, over very elevated swells of undulating prairie, for twenty-five miles, when, on the 9th, we reached the high bluffs of a large tributary called the "Big Witchita river." This stream flows over a clay bed from the southwest and enters Red river about eight miles below Cache creek. It is a deep, sluggish stream, one hundred and thirty feet wide, the water at a high stage very turbid, being heavily charged with red sedimentary matter; the banks abrupt and high, and composed of indurated red clay and dark sandstone. The river is very tortuous in its course, winding from one side to the other of a valley a mile in width, covered with a luxuriant sward of nutritious mezquite grass, which affords the very best pasturage for animals.

The latitude at this place is 34° 25' 51".

There are but few trees on the borders of the Big Witchita: occasionally a small grove of cotton-wood and hackberry is seen; but with this exception, there is no timber or fuel near.

The valley of the river for ten miles above the mouth (the portion I examined) is shut in by bluffs about one hundred feet high, and these are cut up by numerous ravines, in many of which we found springs of pure cold water. The water in the main stream, however, is brackish and unpalatable.

It is my impression that the Big Witchita is of sufficient magnitude to be navigable with small steamers of light draught at almost any stage of water.

In consequence of the high water in Red river, we were detained at the mouth of the Witchita until the morning of the 12th, during which time our provisions being almost consumed, and not knowing positively when our wagon train would join us, I took two Indians with packhorses, swam the river, and started out in quest of it. After going about twenty-five miles towards Fort Arbuckle, we struck the trail of the wagons, and following it two miles, overtook them. They had been detained several days by heavy rains, which had rendered the ground very soft, and in many places almost impassable. In consequence of this, some of the wagons had been broken, and the repairs caused a still further detention. Early on the following morning, after packing the horses with provisions, we returned to where we had left the command, and on our arrival found that the water in the river had fallen sufficiently to admit of fording. Accordingly, on the morning of the 12th, during a violent rain, we commenced the crossing, which was anything but good, as the quicksand in the bed of the river was such as to make it necessary to keep the wagons in constant motion. The moment they stopped, the wheels would sink to the axles, requiring much force to extricate them. By placing a number of men upon each side of the mules and wagons to assist them when necessary, we, however, succeeded in reaching the opposite bank without any serious accident. The latitude at the point where we crossed is 34° 29'. The river is here two hundred yards wide and four feet deep, with a current of three miles per hour; the banks upon each side low and sandy, but not subject to overflow. Passing out through the timbered land on the bottoms, we ascended the high bluff bordering the valley by a gradual slope of about a mile, which brought us upon a very elevated prairie, with the valley of Cache creek in view directly before us. We arrived there on the evening of the 13th, but found that the train had not yet come up. During our march to-day we passed a small stream flowing into Red river, and directly at the point of crossing, in a gully washed out by the rains, we found many pieces of copper ore, of a very rich quality, lying upon the surface.* Our time, however, was too limited to admit of a thorough examination of the locality.

Cache creek is a stream of very considerable magnitude, one hundred and fifty feet wide and three feet deep, with a current of four miles per hour, flowing over a hard clay and gravel bed between high abrupt banks, through a valley one mile in width, of rich black alluvion, and bordered by the best timber I have yet met with west of the Cross Timbers.

---

*An analysis of this ore by Professor Shephard gives the following results:

Copper (with traces of iron)	35.30
Silica	30.60
Oxygen and water	34.10
	100.00

Several varieties of hard wood—such as overcup, pecan, elm, hackberry, ash, and wild china—are found here, among which there is much good timber. The overcup (*Quercus macrocarpa*) especially, is here seen of very unusual size, often from three to four feet in diameter.

This tree, from the length of its stock, the straightness of its grain, and the facility with which it splits, is admirably adapted to building purposes, and is made use of extensively in the southwestern States.

The soil in the valley is of such superior quality, that any kind of grain adapted to this climate could be produced without the aid of irrigation.*

Three miles above the mouth the stream divides into two branches, of about equal magnitude, both of them wooded throughout as far as I traced them, and the soil along them arable in the highest degree; indeed, its fertility is manifest from the very dense and rank vegetation everywhere exhibited. The water in the creek is alkaline, but quite palatable; and its temperature at the time we encamped upon it was 75° F. Our supply train arrived on the 14th; but as the recent rains had raised the water in the creek so much as to prevent our crossing, we were obliged to remain here until the 16th.

This being the point upon Red river at which we were directed to commence our explorations, I propose from this time to make such extracts from my journal as I may conceive pertinent to the objects of the expedition, as set forth in the letter of special instructions, which I had the honor to receive from your office, with such other information as may be considered important, and the conclusions which I have arrived at after an examination of the whole country embraced within the limits of our reconnoissance.

---

*An analysis of the sub-soil from Cache creek, by Professor Shephard, shows that it possesses strong and enduring constituents, and is admirably suited to the production of grain. It is eminently calcareous, as will be seen from the following analysis of its composition:

Silica	82.25
Peroxide of iron	2.65
Alumina	.55
Carbonate of lime	5.40
Carbonate of magnesia	1.70
Water (hygrometric moisture)	5.50
Sulphate of lime and carbonate of potash (only slight traces)	00
	98.05

On the morning of the 16th the water had fallen so much that, after digging down the banks, the wagons were taken over without difficulty. We found an excellent ford upon a rapid where the water was shallow, and the bed hard gravel.

Passing through the timbered land in the bottom, we struck out across the valley, and ascended the ridge dividing Red river from Cache creek; here we found a good road over smooth, high prairie, and after travelling 14.789 miles, encamped upon a small affluent of the west fork of Cache creek, where we found good water and wood. In the course of the march to-day we met with numerous detached pieces of copper ore, mixed with volcanic scoria.* This scoria is found in large masses in the ravines we have passed, and extends back several miles from the creek. The other rocks have been principally sandstone. In the course of the day's march we observed several Indian horse-tracks crossing our road, which were made just previous to the last rain. The direction they had been going was towards the Witchita mountains, and are the first Indian signs we have seen.

---

*These ores consisted of a calcareous amygdaloid, through which is interspersed black oxide of copper and stains of malachite. According to Professor Shephard's analysis, it only yields five per cent. of copper.

Upon the river, a few miles south of our route, we found specimens of a very rich ore, which Professor Shephard, after a careful analysis, pronounces to be a new species, which he has called Marcylite; it was coated with a thin layer of the rare and beautiful Atacamite (muriate of copper,) and consists of—

Copper	54.30
Oxygen and chlorine	36.20
Water	9.50
	100.00

CHAPTER II.

WITCHITA MOUNTAINS—PANTHER KILLED—BUFFALO TRACKS—SINGULAR AND UNACCOUNTABLE RISE OF WATER—BUFFALO SIGNS—HORSE CAPTURED—RAINS—ARRIVAL AT OTTER CREEK—BAROMETER BROKE—CHARACTER OF WITCHITA MOUNTAINS—BUFFALO KILLED—HIGH WATER.

Soon after we had reached the high prairie ridge upon which we travelled to-day, we came in sight of the Witchita mountains, some twenty-five or thirty miles to the north, the chain seeming to be made up of a series of detached peaks, running from the northeast to the southwest as far as the eye can reach. Rising as these mountains do upon the naked prairie, isolated from all other surrounding eminences, they form a very striking and prominent feature in the topography of the country. We cannot yet form any definite estimate as to their height, but shall avail ourselves of the first opportunity to determine this point.

May 17.—On rising this morning I learned, much to my surprise, that nearly all our oxen had wandered off during the night, and had not yet been found. I immediately sent several of the teamsters in search of them; but after being absent two hours they returned unsuccessful, reporting that they could get no track of them. I then started with one of our Delawares, and, after going a short distance from camp, took the track, and following it about a mile came up with the animals, who had very quietly ensconced themselves in a grove of timber near the creek.

As they had upon several occasions before given us trouble, and occasioned the loss of much time, I resolved that in future I would have them herded until late in the evening, and tie them to the wagons for the remainder of the night.

As we did not march until very late this morning, we only made eleven miles, and encamped upon one of the branches of Cache creek.

Our road has continued upon the high ridge lying between Red river and Cache creek, and has been perfectly firm, smooth, and level.

We have to-day seen the first buffalo tracks. They were made during the last rains, and are about five days old. We are anxiously awaiting the time when we shall see the animals themselves, and anticipate much sport.

In the evening, shortly after we had turned out our animals to graze, and had made everything snug and comfortable about us, ourselves reclining very quietly after the fatigue of the day's march, one of the hunters came into camp and informed us that a panther had crossed the creek but a short distance above, and was coming towards us. This piece of intelligence, as may be supposed, created no little excitement in our quiet circle. Everybody was up in an instant, seizing muskets, rifles, or any other weapon that came to hand, and, followed by all the dogs in camp, a very general rush was made towards the spot indicated by the Delaware. On reaching the place, we found where the animal, in stepping from the creek, had left water upon his track, which was not yet dry, showing that he had passed within a short time. We pointed out the track to several of the dogs, and endeavored, by every means which our ingenuity could suggest, to inspire them with some small degree of that enthusiasm which had animated us. We coaxed, cheered, and scolded, put their noses into the track, clapped our hands, pointed in the direction of the trail, hissed, and made use of divers other canine arguments to convince them that there was something of importance on hand; but it was all to no purpose. They did not seem to enter into the spirit of the chase, or to regard the occasion as one in which there was much glory to be derived from following in the footsteps of their illustrious predecessor. On the contrary, the zeal which they manifested in starting out from camp, suddenly abated as soon as their olfactories came in contact with the track, and it was with very great difficulty that we could prevent them from running away. At this moment, however, our old bear-dog came up, and no sooner had he caught a snuff of the atmosphere than, suddenly coming to a stop and raising his head into the air, he sent forth one prolonged note and started off in full cry upon the trail. He led off boldly into the timber, followed by the other dogs, who had now recovered confidence, with the men at their heels, cheering them on and shouting most vociferously, each one anxious to get the first glimpse of the panther. They soon roused him from his lair, and after making a few circuits around the grove, he took to a tree.

I was so fortunate as to reach the spot a little in advance of the party, and gave him a shot which brought him to the ground. The dogs then closed in with him, and others of the party coming up directly afterwards, fired several shots, which took effect and soon placed him "hors du combat." He was a fine specimen of the North American cougar (*Felis concolor,*) measuring eight and a half feet from his nose to the extremity of the tail.

May 18.—At 6 o'clock this morning we resumed our march, taking a course leading to the crest of the "divide," as we thereby avoided many ravines which extended off upon each side towards the streams, and were always sure of a good road for our wagons. This ridge runs very nearly on our course, but occasionally takes us some distance from Red river; as, for example, our encampment of last night was about nine miles from the river, and we only came in sight of it once in the course of our march yesterday.

As soon as the train was under way this morning, Capt. McClellan and myself crossed over the dividing ridge and rode to Red river. We found the bed of the stream about seven hundred yards wide; the valley enclosed with high bluffs upon each side; the soil in the bottom arenaceous, supporting a very spare herbage; and the water very turbid, and spread over a large surface of sand. The general course of the river at this point is a few degrees north of west.

We are all in eager expectation of soon falling in with the buffalo, as we have seen the fresh tracks of quite a large herd to-day. As we advance, the country away from the borders of the water-courses becomes more barren, and woodlands are less frequently met with; indeed, upon the river there is no other timber but cotton-wood (*Populus angulata,*) and elm (*Ulmus Americana,*) and these in very small quantities; for the most part the valley of the river along where we passed to-day is entirely destitute of trees.

We have seen near here several varieties of birds, among which I observed the meadow lark (*Sturnella ludoviciana,*) the pinnated grouse or prairie hen (*Tetrao cupido,*) the Virginia partridge (*Ortyx Virginianus,*) the killdeer (*Charadrius vociferus,*) and several varieties of small birds. We encamped upon a small affluent of Cache creek, where on our arrival we found no water except in occasional pools along the bed; however, in the course of an hour some of the men who had gone a short distance up the creek came running back into camp and crying, at the top of their voices, "Here comes a plenty of water for us, boys!" And, indeed, in a few minutes, much to our astonishment and delight, (as we were doubtful about having a supply,) a perfect torrent came rushing down the dry bed of the rivulet, filling it to the top of the banks, and continued running, turbid and covered with froth, as long as we remained. Our Delawares regarded this as a special favor from the Great Spirit, and looked upon it as a favorable augury to the success of our enterprise. To us it was a most inexplicable phenomenon, as the weather for the last three days had been perfectly dry, with the sky cloudless. If the stream had been of much magnitude we should have supposed that the water came from a distance where there had been rains, but it was very small, extending not more than three miles from the point where we encamped.

Our Delawares report that they have seen numerous fresh buffalo "signs," and that we shall probably soon come upon the herds. We have captured a horse to-day which has a brand upon him, and has probably strayed away from some party of Indians.

May 19.—Last evening the sky became overcast with heavy clouds, and frequent flashes of lightning were observed near the horizon in the north and northwest. Atmospheric phenomena of this character are regarded by the inhabitants of northern Texas as infallible indications of rain, and in verification thereof we had a very severe storm during the night. Much rain has fallen, and the earth has become so soft that I have concluded to remain here until the ground dries a little, particularly as it still continues raining at intervals and the weather is very much unsettled. Frequent rains are very unusual upon the plains at this season of the year; the rainy season generally lasts until about the 1st of May, when the dry season sets in, and there is seldom any more rain until about the middle of August. The past spring has been uncommonly dry—so much so, that vegetation has suffered from it: now, however, the herbage is verdant and the grass most luxuriant.

May 20.—Although it continued raining violently during the night, and the ground was this morning mostly covered with water, we yet made an attempt to travel, but found the prairie so soft that it was with very great difficulty our teams were enabled to drag the wagons over it. We only made five miles and encamped upon a small affluent of Cache creek, which with all the small branches in the vicinity were full to the top of their banks. We find but few trees along the branch upon which we are encamped; hackberry and wild china are the only varieties.

On the 21st we again made an effort to travel; but after going a short distance up the creek, found ourselves obliged, in consequence of the mud, to encamp and await dry weather.

May 22.—This morning, notwithstanding it was cloudy and the ground very far from being dry, we made another effort to proceed. Still keeping the high "divide," we travelled in a westerly direction about eight miles, when we turned north towards two very prominent peaks of the Witchita mountains, and continued in this course until we arrived upon an elevated spot in the prairie, where we suddenly came in sight of Red river, directly before us. Since we had last seen the river it had changed its course almost by a right-angle, and here runs nearly north and south, passing through the chain of mountains in front of us. We continued on for four miles further, when we reached a fine, bold, running creek of good water, which we were all rejoiced to see, as we had found no drinkable water during the day. We encamped about four miles above its confluence with Red river.

This stream, which I have called Otter creek, (as those animals are abundant here,) rises in the Witchita mountains, and runs a course south, 25° west. There are several varieties of wood upon its banks, such as pecan, black-walnut, white ash, elm, hackberry, cotton-wood, wild china, willow, and mezquite; and among these I noticed good building timber. The soil in the valley is a dark loam, and produces a heavy vegetation. The sub-soil is argillaceous. Otter creek is fifty feet wide, and one foot deep at a low stage of water.* The country over which we have passed to-day has been an elevated plateau, totally devoid of timber or water, and the soil very thin and sandy. We have not yet come in sight of any buffaloes, but have seen numerous fresh tracks. Antelopes and deer are very abundant, and we occasionally see turkeys and grouse. Captain McClellan was so unfortunate as to break his mountain barometer last night, which is much to be regretted; as we had brought it so far in safety, we supposed all danger was passed, but by some unforeseen accident it was turned over in his tent and the mercurial tube broken. Fortunately, we have an excellent aneroid barometer, which we have found to correspond very accurately with the other up to this time, and we shall now be obliged to make use of it exclusively.

On ascending Otter creek this morning as high as the point where it debouches from the mountains, I found the timber skirting its banks the entire distance, and increasing in quantity as it nears the mountains. The mountains at the head of the creek have abrupt rugged sides of coarse, soft, flesh-colored granite, mixed with other granulated igneous rocks. Greenstone, quartz, porphyry, and agate are seen in veins running through the rocks, and in some pieces of quartz which were found by Doctor Shumard in the bed of the creek, there were minute particles of gold. As the continued rains have made the ground too soft to admit of travelling at present, we are improving the time by laying in a supply of coal, timber, &c., for our journey on the plains.

May 24.—It commenced raining again during the night, and has continued without cessation all day.

---

*The temperature of the water in the creek at our encampment we found to be 72° F.

May 25.—It has rained violently during all of last night, and has not ceased this morning. When this long storm will abate we do not pretend to form even a conjecture. It has occurred to me that possibly these rains may fall annually in the basin of Upper Red river; thus, perhaps, accounting for what is termed the June Rise in the river. As to the cause of this rise there have been various conjectures; some supposing the river to have its sources in elevated mountain ranges, where the melting of the snows would produce this result; others, again, consider it to be by rains upon the headwaters of the river. This latter idea, however, seems rather improbable, as the country west of the Cross Timbers, so far as known, is generally subjected to very great drought from May to August. We are now in the immediate vicinity of the Witchita mountains, and it is possible they may have an effect upon the weather by condensing the moisture in the atmosphere, and causing rain in this particular locality.

May 26.—Some of the mountains which we ascended yesterday upon the east side of the creek, exhibited a conformation and composition similar to those upon the west side—that of a coarse, soft, flesh-colored granite, the peaks conical, occasionally terminating in sharp points, standing at intervals of from a quarter to one mile apart. In some instances the rocks are thrown together loosely, but here and there showing a very imperfect and irregular stratification, with the seams dipping about twenty degrees with the horizon. The direction of this mountain chain is about south 60° west, and from five to fifteen miles in breadth. Its length we are not yet able to determine. Red river, which passes directly through the western extremity of the chain, is different in character at the mouth of Otter creek from what it is below the junction of the Ke-che-ah-qui-ho-no. There it is only one hundred and twenty yards wide; the banks of red clay are from three to eight feet high, the water extending entirely across the bed, and at this time (a high stage) about six feet deep in the channel, with a rapid current of four miles per hour, highly charged with a dull-red sedimentary matter, and slightly brackish to the taste. Two buffaloes were seen to-day, one of which was killed by our guide, John Bashman.

Deer and antelopes are plenty, but turkeys are becoming scarce as we go west; grouse and quail are also occasionally seen here. As Otter creek continues very high, I intended, if Red river had been fordable, to have crossed that stream this morning and continued up the south bank; but we found the water about eight feet deep, and have no other alternative but to wait until it falls. Along the banks of Red river for the last thirty miles we have observed a range of sand-hills, from ten to thirty feet high, which appear to have been thrown up by the winds, and support a very spare vegetation of weeds, grape-vines, and plum-bushes. Upon the river the timber has diminished so much, that we now find only here and there a few solitary cotton-woods.

From the fact that the Witchita mountains are composed almost entirely of granite and other silicious rocks that usually accompany metallic veins, and that in many places along the range they bear evident marks of great local disturbance, and from the many detached specimens of copper ore found upon the surface throughout this region, I have no doubt but that this will be found, upon examination, to be a very productive mineral district.

CHAPTER III.

WITCHITAS—DISCOURAGING ACCOUNTS OF THE COUNTRY IN ADVANCE—PASS 100° OF LONGITUDE—LEAVE OTTER CREEK—BERRIES—ELK CREEK—PASS WITCHITA MOUNTAINS—GYPSUM BLUFFS—BUFFALOES SEEN—SUYDAM CREEK—COMANCHE SIGNS.

May 27.—As the water still continues at too high a stage for crossing, we moved our camp up the creek about a mile this morning, where we found better grass for our animals. Shortly after we had pitched our tents, a large party of Indians made their appearance on the opposite bank, and requested us to cut a tree for them to cross upon, as they wished to have "a talk" with "the captain." I accordingly had a tall tree cut, which fell across the stream, when they came over upon it and encamped near us.

They proved to be a hunting party of Witchitas, about one hundred and fifty in number, and were commanded by an old chief, "Canaje-Hexie." They had with them a large number of horses and mules, heavily laden with jerked buffalo meat, and ten wild horses which they had lassoed upon the prairie. They said they had been in search of us for several days; having learned we were coming up Red river, they were desirous of knowing what our business was in this part of their country. I replied to them that I was going to the head of Red river, for the purpose of visiting the Indians, cultivating their friendship, and delivering to them "a talk" from the Great Captain of all the whites, who, in token of his kindly feelings, had sent some presents to be distributed among such of his red children as were friends to Americans; and as many of them continue to regard Texas as a separate and independent republic, I endeavored to impress upon them the fact that the inhabitants of that State were of the same nation as the whites in other parts of the United States. I also told them that all the prairie tribes would be held responsible for depredations committed against the people of Texas, as well as elsewhere in our territories. I made inquiries concerning the country through which we still have to pass in our journey.

They said we would find one more stream of good water about two days' travel from here; that we should then leave the mountains, and after that find no more fresh water to the sources of the river. The chief represented the river from where it leaves the mountains as flowing over an elevated flat prairie country, totally destitute of water, wood, or grass, and the only substitute for fuel that could be had was the buffalo "chips." They remarked in the course of the interview that some few of their old men had been to the head of the river, and that the journey could be made in eighteen days by rapid riding; but the accounts given by those who had made the journey were of such a character as to deter others from attempting it. They said we need have no apprehension of encountering Indians, as none ever visited that section of country. I inquired of them if there were not holes in the earth where the water remained after rains. They said no; that the soil was of so porous a nature that it soaked up the water as soon as it fell. I then endeavored to hire one of their old men to accompany me as guide; but they said they were afraid to go into the country, as there was no water, and they were fearful they would perish before they could return. The chief said, in conclusion, that perhaps I might not credit their statements, but that I would have abundant evidence of the truth of their assertions if I ventured much further with my command. This account of the country ahead of us is truly discouraging; and it would seem that we have anything but an agreeable prospect before us. As soon, however, as the creek will admit of fording, I shall, without subjecting the command to too great privations, push forward as far as possible into this most inhospitable and dreaded salt desert. As the Indians, from their own statements, had travelled a great distance to see us, I distributed some presents among them, with a few rations of pork and flour, for which we received their acknowledgments in their customary style—by begging for everything else they saw.

May 28.—Captain McClellan has, by observations upon lunar distances, determined the longitude of our last camp upon the creek to be 100° 0′ 45″, which is but a short distance from the point where the line dividing the Choctaw territory from the State of Texas crosses Red river. The point where this line intersects Otter creek is marked upon a large elm tree standing near the bank, and will be found about four miles from the mouth of the creek upon the south side, with the longitude (100° 0′ 45″) and the latitude (34° 34′ 6″) distinctly marked upon it.

Captain McClellan will start to-morrow morning for the purpose of running the meridian of the 100th degree of longitude to where it intersects Red river, and will mark the point distinctly.

May 29.—After digging down the banks of the creek this morning, we were enabled to cross the train and to resume our march up the river; our course led us towards the point where the river debouches from the mountains, and our present encampment is directly at the base of one of the peaks, near a spring of good water. This mountain is composed of huge masses of loose granite rock, thrown together in such confusion that it is seldom any portion can be seen in its original position. There are veins of quartz, greenstone, and porphyry running through the granite, similar to those that characterize the gold-bearing formation of California, New Mexico, and elsewhere. This fact, in connexion with our having found some small particles of gold in the detritus along the bed of Otter creek, may yet lead to the discovery of important auriferous deposites in these mountains. Among the border settlers of Texas and Arkansas an opinion has for a long time prevailed that gold was abundant here, and several expeditions have been organized among them for the purpose of making examinations, but the Indians have opposed their operations, and in every instance, I believe, compelled them to abandon the enterprise and return home, so that as yet no thorough examination of the mountains has ever been made.*

We find blackberries, raspberries, gooseberries, and currants growing upon the mountains, and this is the only locality west of the Cross Timbers where I have seen them. Grapes and plums are also abundant here, as elsewhere, upon Upper Red river. The grapes are rather smaller than our fox-grapes, are sweet and juicy when ripe, and I have no doubt would make good wine: they grow upon small bushes about the size of currant-bushes, standing erect like them, and are generally found upon the most sandy soil along near the borders of the streams. The plums also grow upon small bushes from two to six feet high, are very large and sweet, and in color vary from a light pink to a deep crimson; they are the Chickasaw plum, (*Prunus chicasa*.)

May 30.—Captain McClellan returned this morning, having traced the meridian of the 100th degree of west longitude to where it strikes Red river. This point he ascertained to be about six miles below the junction of the two principal branches, and three-fourths of a mile below a small creek which puts in from the north upon the left bank, near where the river bends from almost due west to north. At this point a cotton-wood tree standing fifty feet from the water, upon the summit of a sand-hill, is blazed upon four sides facing north, south, east, and west, and upon

---

* Specimens of quartz and black sand were collected in the mountains; and from the presence of hydrated peroxide of iron and iron pyrites in the quartz, and from its similarity to the gold-bearing quartz of California, we were induced to hope that it might contain gold, but a rigid analysis by Professor Shephard did not detect any trace of the precious metal.

these faces will be found the following inscriptions: upon the north side, "Texas, 100° longitude;" upon the south side, "Choctaw Nation, 100° longitude;" upon the east side, "Meridian of 100°, May 29, 1852;" and upon the west side Captain McClellan marked my name, with the date. At the base of the sand-hill will be found four cotton-wood trees, upon one of which is marked "Texas," and upon another will be found inscribed "20 miles from Otter creek."

Red river at this place is a broad, shallow stream, six hundred and fifty yards wide, running over a bed of sand. Its course is nearly due west to the forks, and thence the course of the south branch is WNW. for eight miles, when it turns to nearly NW. The two branches are apparently of about equal magnitude, and between them, at the confluence, is a very high bluff, which can be seen for a long distance around. We are encamped to-night near two mountains about three miles from the river, and one mile west of the head of the west branch of Otter creek near a spring of pure cold water, which rises in the mountains and runs down past our camp. Our road leads along near the creek valley, which is from one to two miles wide, with a very productive soil, covered with a dense coating of grass, and skirted with a variety of hard timber.

May 31.—Our course to-day was northwest until we encountered a bold running stream of good water, forty feet wide and three feet deep, flowing between very high and almost vertical red clay banks, through a broad, flat valley about two miles wide, of a dark alluvial soil, the fertility of which is obvious from the dense vegetation which it supports.

There is a narrow fringe of pecan, elm, hackberry, black walnut, and cotton-wood, along the banks of the creek; but the timber is not so abundant, or of as good a quality, as that upon Otter creek. The abrupt banks made it necessary for us to let our wagons down with ropes. We, however, crossed in a short time, and marched about three miles further, encamping near a small spring of good water, where the wood and grass were abundant.

From the circumstance of having seen elk tracks upon the stream we passed in our march to-day, I have called it "Elk creek." I am informed by our guide that five years since, elk were frequently seen in the Witchita mountains; but now they are seldom met with in this part the country.

The deer and antelopes still continue plenty, but turkeys are scarce. One that our grayhounds caught to-day is the first we have seen for several days. The pinnated grouse, quail, lark, mocking-bird, and swallow-tailed fly-catcher, are also frequently seen.

June 1.—During our march to-day we passed along the borders of a swift running rivulet of clear water which issues from springs in the mountains, and is filled with a multitude of fish. We also passed near the base of a very prominent and symmetrical mountain, which can be seen for twenty miles upon our route, and is a most excellent landmark. Several of the gentlemen ascended this peak with the barometer, and its altitude, as thereby indicated, is seven hundred and eighty feet above the base.

Captain McClellan has called this "Mount Webster," in honor of our great statesman; and upon a rock directly at the summit he has chiselled the names of some of the gentlemen of the party. The valleys lying between many of these mountains have a soil which is arable in the highest degree. They are covered with grasses, which our animals eat greedily. There are also many springs of cold, limpid water bursting out from the granite rocks of the mountains, and flowing down through the valleys, thereby affording us, at all times, a most delicious beverage where we were led to believe, from the representations of the Wichitas, we would find only bitter and unpalatable water. This is an unexpected luxury to us, and we now begin to cherish the hope that all the discouraging accounts of those Indians may prove equally erroneous.

Taking an old Comanche trail this morning, I followed it to a narrow defile in the mountains, which led me up through a very tortuous and rocky gorge, where the well-worn path indicated that it had been travelled for many years. It presented a most wild and romantic appearance as we passed along at the base of cliffs which rose perpendicularly for several hundred feet directly over our heads upon either side. We saw the tracks of several elk that had passed the defile the day previous.

After crossing the mountains, we descended upon the south side, where we found the river flowing directly at the base; and after ascending it about two miles, arrived at a point where it again divided into two nearly equal branches. The water in the south branch (which I have called "Salt Fork") is bitter and unpalatable, and when taken into the stomach produces nausea; whereas that in the other branch, although not entirely free from salts, can be used in cases of great extremity. The compound resulting from the mixture of the water in the two branches below the confluence, is very disagreeable to the taste. The north branch, which I propose to ascend, is, near the junction, one hundred and five feet wide, and three feet deep, with a very rapid current, and the water of much lighter color than that in the Salt Fork. Three miles below the fork, between the river and the base of the mountains, there is a grove of post-oak timber, which Captain McClellan, who examined it, estimates to cover an area of four or five hundred acres. This is well suited for building purposes, being large, tall, and straight. There is also an extensive tract of mezquite woodland near our camp.

One of the Delawares caught two bear cubs in the mountains to-day; one of which he brought in his arms to camp. As the mountain chain crosses the river near here, and runs to the south of our course, we shall leave it to-morrow, and launch out into the prairie before us, following up the bank of the river, which appears to flow through an almost level and uninterrupted plain as far as the eye can extend. I have provided water-casks of sufficient capacity to contain water for the command for three days. I shall always have them filled whenever we find good water; and I hope thereby to be enabled to reach the sources of the river without much suffering. I cannot leave these mountains without a feeling of sincere regret. The beautiful and majestic scenery throughout the whole extent of that portion of the chain we have traversed, with the charming glades lying between them, clothed with a luxuriant sward up to the very bases of the almost perpendicular and rugged sides, with the many springs of delicious water bursting forth from the solid walls of granite, and bounding along over the debris at the base, forcibly reminds me of my own native hills, and the idea of leaving these for the desert plains gives rise to an involuntary feeling of melancholy similar to that I have experienced on leaving home.

June 2.—We left our last night's camp at 3 o'clock this morning, and taking a course nearly due west, emerged from the mountains out into the high level prairie, where we found neither wood nor water until we reached our present position, about half a mile from Red river, upon a small branch with water standing in holes in the bed, and a few small trees scattered along the banks. The latitude at this point is 35° 3'; longitude 100° 12'.

On leaving the vicinity of the mountains, we immediately strike a different geological formation. Instead of the granite, we now find carbonate of lime and gypsum. The soil, except upon the streams, is thin and unproductive. The grass, however, is everywhere luxuriant. Our animals eat it eagerly, and are constantly improving. Near our encampment there are several round, conical-shaped mounds, about fifty feet high, composed of clay and gypsum, which appear to have been formed from a gradual disintegration and washing away of the adjacent earth, leaving the sides exposed in such a manner as to exhibit a very perfect representation of the different strata.

June 3.—We were in motion again at 3 o'clock this morning, our course leading us directly towards a very prominent range of hills situated upon the north bank of Red river, and immediately on the crest of the third terrace or bench bordering the river valley. Their peculiar formation, and very extraordinary regularity, give them the appearance, in the distance, of gigantic fortifications, capped with battlements of white marble. Upon examination they were found to consist of a basis of green or blue clay, with two super-strata of beautiful snow-white gypsum, from five to fifteen feet in thickness, resting horizontally upon a sub-stratum of red clay, with the edges wholly exposed, and so perfectly symmetrical that one can with difficulty divest himself of the idea that it must be the work of art, so much does it resemble masonry. In many places there are perfect representations of the re-entering angles of a bastion front, with the glacis revetted with turf, and sloping gently to the river. Several springs issue from the bluffs, and (as I have always found it to be the case in the gypsum formation) the water is very bitter and disagreeable to the taste.

I am inclined to believe that this same formation extends in a south-westerly direction from the Canadian river to this place, as I passed through a belt of country upon that stream somewhat similar to this, and in a position to be a continuation of it. We crossed the river near the lower extremity of the bluffs at a point where we found it fifty yards wide and sixteen inches deep, with a current of three miles per hour, running over a bed of quick-sand. We passed without difficulty by keeping the animals in rapid motion while in the stream, and encamped upon the high bluff on the south side. By following up the course of a ravine in the side of the gypsum bluffs, where there were detached pieces of copper ore, we discovered a vein of this metal which proved to be the "green carbonate," but not of so rich a character as that we had seen before. At this point we are nearly opposite the western extremity of the chain of Witchita mountains.*

---

*Professor Shephard's analysis of a specimen of the sub-soil from the valley of the river near our camp on the third June, gives the following result:

Silica	79.30
Peroxide of iron	8.95
Alumina	1.50
Carbonate of lime	1.10
Sulphate of lime, with strong traces of sulphate of soda and chloride of sodium	4.65
Water	4.50
	100.00

June 4.—We made an early start this morning, and travelled in the direction of a chain of bluffs which appeared to us to be upon the branch of the river we were ascending; but on reaching them we found ourselves upon a creek running towards the *Salt Fork*, the bluffs of which we could see from the top of an eminence near the creek, about eight miles distant.

To regain our route we were obliged to turn directly north, and march about six miles in this direction, when we again came in sight of the main *North Fork*. In our route we have passed near several hills of similar formation to that of the gypsum bluffs before described. Sulphate of lime is found in large quantities throughout this section, and occurs in various degrees of purity, from the common plaster of Paris to the most beautifully transparent selenite I have ever seen. I observed several specimens, from one to two inches in thickness, that were as absolutely colorless and limpid as pure water.

We are encamped upon the elevated prairie near a clump of trees, where we find water standing in pools. We have found the grass abundant, and the water and wood sufficiently so for our purposes at all our camps since we left our visitors, the Witchitas.

As I was riding to-day with one of our Delawares, about three miles in advance of the train, we suddenly (as we rose upon an eminence in the prairie) came in sight of four buffalo cows with calves, very quietly grazing in a valley below us. We at once put spurs to our horses, and, with our rifles in readiness, set out at a brisk gallop in pursuit; but, unfortunately, they had "the wind" of us, and were instantly bounding off over the hills at full speed. We followed them about three miles, but as they were much in advance at the outset we could not overtake them without giving our horses more labor than we cared about, and so abandoned the chase. Our grayhounds caught two young deer upon the open prairie to-day, and they have had several chases in pursuit of the antelope, but have not as yet been able to come up with them. The latitude of our present position is 35° 15' 43".

June 5.—After marching nearly a mile from our last camp, we crossed a running brook of clear water, which had a slightly sulphurous taste and odor. It rises in the hills to the southwest and runs rapidly, like a mountain stream, into the main river. The appearance of this stream reminded me so forcibly of some I have seen in the mountains of Pennsylvania, that I searched it faithfully, expecting to see the spotted trout, but only found a few sun-fish and minnows.

From this brook to our present position, the country we traversed was exceedingly monotonous and uninteresting, being a continuous succession of barren sand-hills, producing no other herbage than the artemisia, and a dense growth of dwarf oak bushes, about eighteen inches high, which seem to have attained their full maturity, and bear an abundance of small acorns. The same bush is frequently met with upon the Canadian river, near this longitude, and is always found upon a very sandy soil. Our camp is in the river valley near a large spring of sulphurous water, in the midst of a grove of cotton-wood trees. Upon a creek we passed to-day on the opposite bank of the river we noticed pecan, elm, hackberry, and cotton-wood trees. The grass still continues good, and the water of the main river, although not good, can be used. The bed of the river is here one hundred yards wide, with but little water passing over the surface, being mostly absorbed by the quick-sands. Our Indians brought in three deer this evening, and the grayhounds have caught a full-grown doe in a fair chase upon the open prairie. We occasionally see a few turkeys, but they are not as abundant as we found them below here. There are several varieties of birds around our camp—among which we saw the white owl, meadow-lark, mocking-bird, king-bird, swallow, swallow-tailed fly-catcher, and quail.

June 6.—Starting at 3 o'clock this morning, we crossed the river near our last camp, and passed over a very elevated and undulating prairie for ten miles, when we reached a large creek flowing into Red river, which, in compliment to my friend Mr. J. R. Suydam, of New York city, who accompanied the expedition, I have called "Suydam creek." It is thirty feet wide; the water clear, but slightly brackish, and flows rapidly over a sandy bed between abrupt clay banks, which are fringed with cotton-wood trees. As the water in the main river near our camp is very bitter, we were obliged to make use of that in the creek.

Above our present encampment there appears to be a range of sand-hills, about three miles wide, upon each side of the river, which are covered with the same herbage as those we passed below here.

We have seen the trail of a large party of Comanches, which our guide says passed here two days since, going south. I regret that we did not encounter them, as I was anxious to make inquiries concerning our onward route. These Indians were travelling with their families. Upon a war expedition they leave their families behind, and never carry lodges, encumbering themselves with as little baggage as possible. On the other hand, when they travel with their families, they always carry all their worldly effects, including their portable lodges, wherever they go; and as they seldom find an encampment upon the prairies where poles for the frame-work of the lodges can be procured, they invariably transport them from place to place, by attaching them to each side of the pack-horses, with one end trailing upon the ground. These leave parallel marks upon the soft earth after they have passed, and enable one at once to determine whether the trail is made by a war party or otherwise. The Comanches, during the past year, have not been friendly with the Delawares and Shawnees; and although there has as yet been no organized demonstration of hostilities, they have secretly killed several men, and in consequence our hunters entertain a feeling of revenge towards them. They, however, go out alone every day upon their hunts, are frequently six or eight miles from the command, and seem to have no fears of the Comanches, as they are liable to encounter them at any moment; and being so poorly mounted that they could not escape, their only alternative would be to act on the defensive. I have cautioned them upon the subject several times, but they say that they are not afraid to meet any of the prairie Indians, provided the odds are not greater than six to one. They are well armed with good rifles—the use of which they understand perfectly—are intelligent, active, and brave, and in my opinion will ere long take ample satisfaction upon the Comanches for every one of their nation that falls by their hands.

CHAPTER IV.

BUFFALO CHASE—SWEET WATER CREEK—COMANCHE CAMPS—PREVAILING WINDS—INDIANS SEEN—METHOD OF ENCAMPING—WONDERFUL POWERS OF THE DELAWARES—BEAVER DAMS—KIOWAY CREEK.

June 7.—Taking two of the Indians this morning, I went out for the purpose of making an examination of the surrounding country and ascertaining whether good water could be found upon our route for our next encampment. We had gone about three miles in a westerly direction, when we struck a fresh buffalo track leading north; thinking we might overtake him, we followed up the trace until we came near the summit of an eminence upon the prairie, when I sent one of the Indians (John Bull) to the top of the hill, which was about one-fourth of a mile distant, to look for the animal. He had no sooner arrived at the point indicated than we saw him make a signal for us to join him, by riding around rapidly several times in a circle and immediately putting off at full speed over the hills. We set out at the same instant upon a smart gallop, and on reaching the crest of the hill discovered the terrified animal fleeing at a most furious pace, with John Bull in hot pursuit about five hundred yards behind him. As we followed on down the prairie we had a fine view of the chase. The Delaware was mounted upon one of our most fractious and spirited horses, that had never seen a buffalo before, and on coming near the animal he seemed perfectly frantic with fear, making several desperate surges to the right and left, any one of which must have inevitably unseated his rider had he not been a most expert and skilful horseman. During the time the horse was plunging and making such efforts to escape, John, while he controlled him with masterly adroitness, seized an opportunity and gave the buffalo the contents of his rifle, breaking one of his fore-legs, and somewhat retarding his speed: he still kept on, however, making good running, and it required all the strength of our horses to bring us alongside of him. Before we came up our most excellent hunter, John Bull, had recharged his rifle and placed another ball directly back of the shoulder; but so tenacious of his life is this animal, that it was not until the other Delaware and myself arrived and gave him four additional shots, that we brought him to the ground. Packing the best pieces of the meat upon our horses, we went on, and in a few miles found a spring-brook, in which there was an abundance of good water, where I determined to make our next encampment. On our return we saw a pack of wolves, with a multitude of ravens, making merry over the carcass of the buffalo we had killed in the morning.

Thinking that the Comanches, whose trail we had seen yesterday, might possibly be encamped within a few miles of us, I this morning directed Captain McClellan to take the interpreter and follow the trace. After going about fifteen miles he found one of their camps that had been abandoned two days previous; and as there was no prospect of overtaking them he returned, after ascertaining that they were travelling a southerly course towards the Brazos river.

In many places above the Witchita mountains we have found drift of quartz and scoria, but the boulders of greenstone, granite, and porphyry, were only seen below the upper end of the range; and the nearer we approached the mountains from below, the larger and more angular became the fragments, until, on reaching near the base, large angular pieces nearly covered the surface of the ground, thereby leading us to the conclusion that here is the source of the boulders we have seen below the mountains, whereas the drift found here must come from above, as we have yet discovered no igneous rocks in place since we left the mountains. The formation here is a dark limestone overlaid with loose scoria. The earth upon the stream is highly arenaceous, and the soil poor. The grass, however, as we have found it everywhere upon Red river and its tributaries, is of a very superior quality, consisting of several varieties of grama and mezquite.

The range of the grama grass, so far as my observations have extended, is bounded on the north by near the parallel of 36° north latitude, and on the east by about the meridian of 98° west longitude. It extends south and west, as far as I have travelled; it appears, however, to flourish better in about the latitude of 33° than in any other. As there is generally a drought on these prairies from about the 1st of May to the middle of August, it would appear that the particular varieties of grasses that grow here do not require much moisture to sustain them.

June 8.—Our route to-day has been over a rolling prairie, in many places covered with the dwarf oak bushes before mentioned. We are encamped upon a creek of clear and wholesome water, which Dr. Shumard has named "Loess creek," from the circumstance that the soil upon the stream contains a deposite of land and fresh-water shells, among which are found those of *Pupa muscorum, Succinea elongata,* and *Helix plebeium,* forming a pulverent grayish loam similar to the loess found upon the Rhine.

No fossils were seen in this silt, but our time would not admit of making a very thorough examination of the locality. Specimens of the shells were, however, procured, to accompany our collection, and were found to be similar to those described by Lyell as occurring in Europe.

The creek is twenty feet wide and eight inches deep; runs rapidly between low banks, with only a few cotton-wood and elm trees upon them. There are also some few small knots or clumps of trees upon the elevated prairie lands in the vicinity. The observations for latitude at this point give the result 35° 24' 50".

June 9.—At half-past 2 o'clock this morning we were *en route* again over a very elevated prairie for six miles, when we arrived in the valley of a fine stream of pure water, twelve feet wide and one foot deep, with a rapid current. This stream is fringed by large cotton-wood trees along the banks, and the grass in the valley is most excellent, consisting of the mesquite and wild rye, which our animals are very fond of. From the fact of the water being so good in this stream, we called it Sweet-water creek. The valley is bordered upon each side by bluffs from ten to forty feet high; the soil a reddish loam and quite productive, being somewhat similar in appearance to that in the bottoms of Red river below the confluence of the Witchita, where the most abundant crops are produced.

As we ascend the river we have conclusive evidence of the falsity of the representations of our visitors, the Witchitas. It will be remembered they told us that the entire country was a perfectly desolate waste, where neither man nor beast could get subsistence, and that there was *no danger from Indians,* as none ever resorted to this section of Red river. Their statements have proved false in every particular, as we have thus far found the country well watered, the soil in many places good, everywhere yielding an abundance of the most nutritious grasses, with a great sufficiency of wood for all the purposes of the traveller.

There are several old camps near us, which appear to have been occupied some two or three weeks since by the Comanches: the grass where their animals have grazed is not yet grown up.

Red river, which is about six miles distant from our present position, is eighty yards wide, with but a very small portion covered with water, running over the quick-sand bed. The banks upon each side are from four to ten feet high, and not subject to inundation. The valley is here about half a mile wide, shut in by sandy bluffs thirty feet high, which form the border to a range of sand-hills extending back about five miles upon each side of the river. The soil in the valley is sandy and sterile, producing little but scattering weeds and stunted brush.

June 10.—Our course to-day has been almost due west, up the north bank of Sweet-water creek. The country upon each side of the valley is high and gently undulating, and the geological formation has changed from deep-red sandstone to carboniferous limestone.

The weather for the last four days has been very cold, as will be seen from the meteorological tables appended; indeed, I think I have never in this latitude known the thermometer to range as low at this season. Upon the plains where I have heretofore travelled during the summer months, a strong breeze has generally sprung up about 8 o'clock in the morning and lasted until after night, reaching its maximum intensity about 3 o'clock in the afternoon. This breeze comes from the south, and generally rises and subsides with as much regularity as the sea-breeze upon the Atlantic coast, which fact has given rise to the opinion that it comes from the Gulf of Mexico. These cool and bracing winds temper the atmosphere, heated to intensity by the almost vertical rays of the sun, rendering it comfortable and even pleasant in midsummer. Observations were made this evening for the determination of latitude, and the result showed 35° 26' 13."

June 11.—We crossed Sweet-water creek at 3 o'clock this morning, and, keeping back upon the high prairie bordering the valley, travelled eight miles in nearly a west course, when we crossed two fresh Indian trails, which, from the circumstance of there being no trace of lodge-poles, our guide pronounced to have been made by war parties; and he states that he has during the day seen four Indians upon a hill in the distance taking a look at us, but that they turned immediately on seeing him and galloped off. The fact of their not being disposed to communicate with us looks suspicious, and they may have hostile intentions towards us; but with our customary precautions, I think we shall be ready to receive them, either as friends or enemies.

Our usual method of encamping is, where we can find the curve of a creek (which has generally been the case,) to place ourselves in the concavity, with the wagons and tents extending around in a semi-circle, uniting at each extremity of the curve of the creek so as to enclose a sufficient space for the command; thus we are protected on one side by the creek, and upon the other by the line of wagons and tents. Immediately after reaching our camping-ground, all the animals are turned out to graze, under charge of the teamsters, who are armed and remain constantly with them, keeping them as near the command as the supply of grass will permit. We generally commence the day's march about 3 o'clock in the morning, and are ready to encamp by 11 o'clock; this gives ample time for the animals to graze before night, when they are driven into camp. The horses and mules are picketed within the enclosure, while the oxen are tied' up to the wagons; sentinels are then posted upon each side of the encampment, and kept constantly walking in such directions that they may have the animals continually in view.

Many have supposed that cattle in a journey upon the plains would perform better and keep in better condition by allowing them to graze in the morning before starting upon the day's march, which would involve the necessity of travelling during the heat of the day. These persons are of opinion that animals will only feed at particular hours of the day, and that the remainder of the day must be allotted them for rest and sleep, and that unless these rules are adhered to they will not thrive. This opinion, however, is, I think, erroneous, and I also think that cattle will adapt themselves to any circumstances so far as regards their working hours and their hours of rest. If they have been accustomed to labor at particular hours of the day, and the order of things is at once reversed, the working hours being changed into hours of rest, they may not do as well for a few days, but they soon become accustomed to the change, and eat and rest as well as before.

By starting at an early hour in the morning during the summer months, the day's march is over before it becomes very warm; whereas, (as I have observed) if the animals are allowed time to graze before starting, the march must continue during the middle of the day, when the animals (particularly oxen) will suffer much from the heat of the sun, and, so far as my experience goes, will not keep in as good condition as when the other plan is pursued. I have adopted this course from the commencement of our journey, and our oxen have continued to improve upon it. Another and very important advantage to be derived from this course is found in the fact that the animals, being tied up during the night, are not liable to be lost or stolen.

The country over which we are now passing, except directly in the valleys of the streams, is very elevated and undulating, interspersed with round conical hills, thrown up by the winds, with the apices very acute; the soil, a light gray sand, producing little other vegetation than weeds and dwarf oaks.

The creek up which we have been travelling runs almost parallel to Red river, and affords us fine camping-places at any point.

From the very many old Indian camps that we have seen, and the numerous stumps of trees which at different periods have been cut by the Indians along the whole course of the creek, we infer that this is, and has been for many years, a place of frequent resort for the Comanches, and I have no doubt they could always be found here at the time the buffaloes are passing back and forth in their migrations during the spring and winter.

The parties of Indians whose trails we crossed in our march to-day were going south, and not having their families with them, our interpreter infers that they are bound for Mexico upon a foray. Had we met them and learned that such was their intention, we might perhaps have dissuaded them from proceeding further. They may have seen our trail: if so, and they are friendly, they will visit us. Should they not come in, however, I shall send out an Indian after them to ascertain where they encamped and the time they left. In consequence of their known hostility, our Delawares are getting somewhat cautious about encountering them. The interpreter says he would not be afraid to meet five or six, but thinks he would avoid a greater number. I directed him, in the event of his meeting a party, to invite them to come to camp, as I had a talk for them. He replied, "Suppose he want to kill me—I not tell him."

This man has often been among the prairie Indians, understands their language and character well, and the moment he sees a trail made by them, or an old deserted camp, he at once determines of what nation they were; the number of horses and mules in their possession; whether they were accompanied by their families, and whether they were upon a war expedition or otherwise; as also the time (within a few hours) of their passing, with many other facts of importance.

These faculties appear to be intuitive, and confined exclusively to the Indian: I have never seen a white man that could judge of these matters with such certainty as they. For example, upon passing the trail of the Indians to-day, one of our Delawares looked for a moment at the foot-prints, picked up a blade of grass that had been crushed, and said the trail was made two days since, when to us it had every appearance of being quite fresh; subsequent observations satisfied us that he was correct.

Upon another occasion, in riding along over the prairie, I saw in the sand what appeared to me to be a bear-track, with the impression of all of the toes, foot, and heel; on pointing it out to one of the Indians, he instantly called my attention to some blades of grass hanging about ten inches over the marks, and explained to me that while the wind is blowing, these blades are pressed towards the earth, and the oscillation thereby produced had scooped out the light sand into the form I have mentioned. This, when explained, was perfectly simple and intelligible; but I am very much inclined to believe the solution of it would have puzzled the philosophy of a white man for a long time.

A few such men as the Delawares attached to each company of troops upon the Indian frontier would, by their knowledge of Indian character and habits, and their wonderful powers of judging of country, following tracks, &c., (which soldiers cannot be taught) enable us to operate to much better advantage against the prairie tribes. In several instances when we have had our animals stray away from camp, I have sent six or eight teamsters for them, who, after searching a long time, would often return unsuccessful. I would then send out one Indian, who would make a circuit around the camp until he struck the tracks of the lost animals, and following them up, would invariably return with them in a short time. In this way their services are almost indispensable upon an expedition like ours.

June 12.—Our course to-day was very nearly due west, up the left bank of Sweet-water creek, until, within about three miles of our present position, we turned with the course of the stream more northwardly.

The country we passed over was similar to that of yesterday, but not so sandy or so heavy upon our teams. We came in sight of a line of high bluffs this morning, which were apparently about ten miles to the northwest of us. They are very elevated, and present much the appearance of the borders of the great Staked Plain, or the "Llano estacado" of the Mexicans.

On reaching camp we found that a large party of Indians, with very many animals, had been encamped here about two weeks since. Numerous trails and horse tracks were seen in every direction, and their animals have cropped the grass for a long distance around.

Their lodges were pitched near our camp, and our guide pronounced them to have been Kioways. On inquiring how he could distinguish a Kioway from a Comanche camp, he said the only difference was that the former make the holes for their fires about two feet in diameter, while the latter only make them about fifteen inches.

A community of beavers have also selected a spot upon the creek near our camp, for their interesting labors and habitations. I know of no animal concerning which the accounts of travellers have been more extraordinary, more marvellous or contradictory, than those given of the beaver. By some he is elevated in point of intellect almost to a level with man. He has been said, for instance, to construct houses, with several floors and rooms; to plaster the rooms with mud in such a manner as to make smooth walls, and to drive stakes of six or eight inches in diameter into the ground, and to perform many other astounding feats, which I am inclined to believe are not supported by credible testimony. Laying aside these questionable statements, there is quite sufficient in the natural history of the beaver to excite our wonder and admiration. For instance, at this place, upon an examination of the dam they have constructed, I was both astonished and delighted at the wonderful sagacity, skill, and perseverance which they have displayed. In the selection of a suitable sight, and in the erection of the structure, they appear to have been guided by something more than mere animal instinct, and have exhibited as correct a knowledge of hydrostatics, and the action of forces resulting from currents of water, as the most scientific millwright would have done. Having chosen a spot where the banks on each side of the creek were narrow and sufficiently high to raise a head of about five feet, they selected two cotton-wood trees about fifteen inches in diameter, situated above this point, and having an inclination towards the stream: these they cut down with their teeth, (as the marks upon the stumps plainly showed,) and, floating them down to the position chosen for the dam, they were placed across the stream with an inclination downward, uniting in the centre. This formed the foundation upon which the superstructure of brush and earth was placed, in precisely the same manner as a brush dam is made by our millwrights, with the bushes and earth alternating and packed closely, the butts in all cases turned down the stream. After this is raised to a sufficient height, the top is covered with earth, except in the centre, where there is a sluice or waste-wier, which lets off the superfluous water when it rises so high as to endanger the structure. In examining the results of the labors of these ingenious quadrupeds, it occurred to me that the plan of erecting our brush dams must have been originally suggested from witnessing those of the beavers, as they are very similar. I watched for some time upon the banks of the pond, but could see none of the animals. I presume they think we make too much noise in our camp to suit them, and deem it most prudent to remain concealed in their sub-marine houses.

I observed one place above the pond where they had commenced another dam, and had progressed so far as to cut down two trees on opposite sides of the creek; but as they did not fall in the right direction to suit their purposes, the work was abandoned. As the course of Sweet-water creek turns too much to the north above here, we shall leave it; and it is with much regret that we are obliged to do so, as it has afforded us the best of spring water, with good grass and wood, for five days.

June 13.—Leaving the command this morning encamped upon Sweet-water creek, I made a trip to Red river, which is about six miles in a southwest direction; it was one hundred yards wide where we struck it, with but a very small portion covered with water, and, very much to our astonishment, for the first time, upon tasting it, we found it free from salts. Following up the stream about a mile, we discovered that this good water all issued from a small stream that put in upon the north bank, and above this the bed of the main river was dry. As there is an incrustation of salt upon the bed of the river below the creek, where the water has subsided after a high stage, I have no doubt but that the water above here will be found to be impregnated with salts, and that all the fresh water now found in the river comes from the creek mentioned.

Along the whole course of Red river, from Cache creek to this point, we find three separate banks or terraces bordering the river; the first of which rises from two to six feet above the bed of the stream. The second is from ten to twenty feet high; and the third, which forms the high bluff bordering the valley of the river, is from fifty to one hundred feet. The first bank is in places subject to inundation, and generally is from fifty to two hundred yards wide. The second is never submerged, and is from two to fifteen hundred yards wide. The third bank bounds the high prairie. We found the range of sand-hills still continuing along the river; and we have constantly during the day been in sight of the line of bluffs which I supposed to be the border of the "Llano estacado." We also passed the trail of a very large party of Indians, who were ascending the river before the last rain, (some two weeks since.)

After leaving the river on our return to camp, we found two fine brooks of cold spring water, with good wood and grass upon them, and as they are in our course I propose to make our next camp upon one of them.

June 14.—Making an early start this morning, we travelled eleven miles in a westerly course, when we reached a very beautiful stream of good spring water, flowing with a uniformly rapid current through a valley about a mile wide, covered with excellent grass. There is a heavy growth of young cotton-wood trees along the borders of the creek, and among them are found immense quantities of that peculiar variety of grape I have before mentioned as growing in the sand-hills along the valley of Red river. They grow here upon low bushes about four feet high, similar to those cultivated varieties that are trimmed and cut down in the spring. When growing near the trees they never rest upon them, like our eastern varieties of the wild grape, but stand separate and erect, like a currant-bush.

This creek appears to be a place of winter resort for large numbers of the prairie Indians. We found many old camps along the stream, and the ground for several miles was thickly strewn with cotton-wood sticks, the bark of which had been eaten off by their animals. The prairie tribes are in the habit of feeding their favorite horses with the cotton-wood bark in the winter; and it is probably the abundance of this wood that has attracted them here. We found the stumps of the trees they had cut from year to year in various stages of decay—some entirely rotten, and others that had been cut during the past winter. The fine mezquite and grama grass furnishes pasturage for their animals during a great part of the winter; and the cotton-wood is a never-failing resort when the grass is gone.

As we are now nearly opposite the country on the Canadian river occupied by the Kioway Indians, it is quite probable that some of that nation winter at this place; and I have no doubt but that they could be found here at any time during that season. I have called the creek *Kioway creek*.

Game is abundant in this vicinity; and our hunters keep the entire command constantly supplied with fresh meat, so that we have not yet had occasion to kill one of our beef-cattle. Seven deer and one antelope were killed to-day. For months previous to leaving Fort Belknap, with the exception of a few wild onions, my men had eaten no vegetables. Some of them had been attacked with scurvy, and all were more or less predisposed to it. I have, therefore, been exceedingly anxious to take all possible precautions for warding off this most dreaded disease. As I had no anti-scorbutic, with the exception of a very few dried apples and a little citric acid, I was obliged to make use of everything the country afforded as a substitute for vegetables. I caused the men to eat greens whenever they could be obtained, with the green grapes occasionally; and to-day we were so fortunate as to discover a fine bed of wild onions (a most excellent anti-scorbutic) upon some sand-hills over which we passed. A quantity were collected by the men and made use of freely.

CHAPTER V.

REACH THE SOURCE OF THE NORTH BRANCH OF RED RIVER—BOTTLE BURIED—ARRIVED UPON THE CANADIAN—DEPARTURE FOR MIDDLE FORK—INDIAN BATTLE-GROUND—PRAIRIE-DOG TOWNS—SOURCE OF THE MIDDLE FORK—SOUTH FORK—PRAIRIE-DOGS.

June 15.—On account of the morning being dark and the clouds threatening rain, we did not leave camp until daylight this morning. We, however, made a good day's march over a very heavy sandy country, and after crossing the main river, encamped upon the south bank.

During the day we crossed several small branches, in which we found good water; and in several places where there was timber upon them, we saw old Indian camps. At one place I noticed a large grove of cotton-wood which had been entirely enclosed with a brush fence by the Indians: this was probably made for the purpose of keeping their animals from straying away.

On reaching the river we found that it had very much diminished in magnitude since we had last seen it. It was now only fifteen yards wide, the water clear, and to the taste entirely free from salts.

The herbage for the last twenty miles of our march has suffered much from drought, and the grass in many places upon the elevated lands is entirely burnt up. We, however, continue to find excellent grass in the valleys near the borders of the small streams, and upon the river itself. The only varieties of timber that we find upon this part of Red river are cotton-wood and hackberry, the former greatly predominating and of large dimensions. Indeed, I have never seen so much timber at any other place upon the plains, in this longitude, as we find here.

We have had the line of high bluffs in sight before us all day, and we are now within a few miles of them. The geological formation through the country over which we are passing is a light-colored calcareous sandstone, covered with a drift of quartz and scoria.

Near our present position, upon the opposite side of the river, there has been a very large band of Kioways encamped, about two weeks since, and their animals have cropped much of the grass for several miles around us. From the multitude of tracks that we see in every direction, there must have been an immense number of animals. On leaving here their course was south.

June 16.—Striking our tents at three o'clock this morning, we followed up the south bank of the river, which runs in a westerly course for eight miles, when it suddenly turns to the southwest, and here the elevated bluffs which we have had in view for several days past approach the river upon each side, until there is but a narrow gorge or cañon for the passage of the stream. These bluffs are composed of calcareous sandstone and clay, rising precipitously from the banks of the stream to the height of three hundred feet, when they suddenly terminate in the almost perfectly level plain of the "Llano estacado." Here the river branches out into numerous ramifications, all running into the deep gorges of the plain. Taking the largest, we continued up it, riding directly in the bed of the stream for about five miles, when we reached the source of this branch of the river; and by ascending upon the tablelands above, we could see the heads of the other branches which we had passed a few miles below.

The latitude at this place, as determined by several observations of Polaris, is 35° 35' 3", and the longitude 101° 55'. These results make our position only about twenty-five miles from the Canadian river; and as I am anxious to determine how our observations conform to those we made in ascending that stream in 1849, I propose taking ten men, and leaving the main body of the command to guard our oxen and stores, to make a trip in a due north course to the Canadian. This will serve to show the connexion between that stream and a certain known point upon the head of the north branch of Red river; and is, in my opinion, a geographical item which it is important to establish and confirm by actual observation, particularly as the Canadian has by several travellers been mistaken for Red river.

At our encampment of this evening is the last running-water we have found in ascending this branch of Red river. We are near the junction of the last branch of any magnitude that enters the river from the north, and about three miles from the point where it debouches from the plains, in a grove of large cotton-wood trees upon the south bank of the river. Under the roots of one of the largest of these trees, which stands near the river, and below all others in the grove, I have buried a bottle containing the following memorandum: "On the 16th day of June, 1852, an exploring expedition, composed of Captain R. B. Marcy, Captain G. B. McClellan, Lieutenant J. Updegraff, and Doctor G. C. Shumard, with fifty-five men of company D fifth infantry, encamped here, having this day traced the north branch of Red river to its sources. Accompanying the expedition were Captain J. H. Strain, of Fort Washita, and Mr. J. R. Suydam, of New York city." This tree is blazed on the north and east sides, and marked upon the north side with a pencil as follows: "Exploring Expedition, June 16, 1852."

An incident happened this evening, which for a short time gave us much uneasiness and alarm. It was caused by one of the gentlemen of the party walking out from camp alone without our knowledge, and remaining away about two hours before we discovered his absence. It was after dark when I first learned that he was not in camp; and as there were many fresh signs of Indians around, I was fearful he had fallen into their hands. I immediately started out the Delawares in search of him, and ordered our six-pounder to be discharged, with muskets at short intervals, and at the same time made preparations for starting out myself; but no sooner had the cannon been fired than he made his appearance, in a state of much excitement, and had evidently been greatly confused and alarmed, as is always the case with persons who are lost. He states that he had gone out for the purpose of taking a short walk, and in returning over a hill had lost sight of the camp; that in endeavoring to make his way back he had become so much confused, that after night he took ours for a Comanche camp, and dared not approach until he heard the signal-gun.

June 17 to 19.—On the 17th, accompanied by three gentlemen of the party, with five soldiers and three Indians, I started in a northerly direction to go in search of the Canadian river. Our route led us immediately out upon the elevated plateau of the Staked Plain, where the eye rests upon no object of relief within the scope of vision.

Pursuing our way over this monotonous and apparently boundless plain for fifteen miles, our eyes were suddenly gladdened by the appearance of a valley and bluffs before us, which I at once recognised to be upon the Canadian; and after travelling ten miles further, we found ourselves upon that stream, making the entire distance from the head of Red river to the Canadian twenty-five miles. This was a matter of much gratification and interest to us, as it developed and confirmed the accuracy of our calculations regarding the geographical position of the sources of Red river. The point where we struck the Canadian is at the mouth of a small stream called Sandy creek upon the map of the road from Fort Smith to Santa Fé. This being near longitude 101° 45', and latitude 35° 58', makes the calculations for the two positions approximate very closely. The formation upon the Canadian at this point is very similar to that upon the Red river, being composed of light-colored friable arenaceous limestone, resting upon a stratum of red sand, with a sub-stratum of blue clay; the whole overlaid by a drift of quartz, felspar, and agate. The soil upon the creek is a dark-brown loam, covered with a heavy coating of wild rye and mezquite; and if the drought of summer did not prevent, would produce abundant crops. The only varieties of timber found here are the wild china, hackberry, willow, and cotton-wood; the latter, in some instances, growing to an enormous size. One tree, standing upon the creek near the Canadian, which we measured, was nineteen and a half feet in circumference at five feet above the ground. The Santa Fé road passes directly along the river-bank at this place, and upon the north side of the river stand four cotton-wood trees; these are blazed, and the distance in a due south course to the head of Red river, with the date of our arrival there, marked upon one of them. Having finished the examination of the north branch of Red river, we propose turning to the south from this point, and, crossing the elevated prairie of the Staked Plain, shall endeavor to reach the middle or Salt Fork, which we passed upon our left near the upper extremity of the Witchita range of mountains. The only apprehension we entertain is, that we may suffer for water, but shall keep our water-casks filled whenever it is practicable.

The grass upon the Staked Plain is generally a very short variety of mezquite, called buffalo-grass, from one to two inches in length, and gives the plains the appearance of an interminable meadow that has been recently mown very close to the earth.

I have never travelled over a route on the plains west of the Cross Timbers where the water, grass, and wood were as good and abundant as upon the one over which our explorations have led us. This has been to us a most agreeable surprise, as our friends, the Witchitas, had given us to understand that we should find no wood, and nothing but salt water, in this section of country. I can account for their misrepresentations only on the ground that they did not wish us to go into the country, and took this course to deter us from proceeding further.

June 20.—We made an early march this morning, passing over the high hills bordering the river, and the broad swells of prairie adjoining, for twelve miles, when we reached the valley of a very beautiful stream, twenty feet wide and six inches deep, running rapidly over a gravelly bed, through a valley about a mile wide of sandy soil, with large cotton-wood trees along the banks. I have called this "McClellan's creek," in compliment to my friend Captain McClellan, who I believe to be the first white man that ever set eyes upon it.

We were happy, on arriving here, to find the water perfectly pure and palatable; and we regard ourselves as most singularly fortunate in having favorable weather. The rains of the last two days have made the atmosphere delightfully cool, and afford us water in many places where we had no reason to expect it at this season of the year.

During the middle of the day, when the earth and the adjacent strata of air had become heated by the almost vertical rays of the sun, we observed, as usual, upon the "Llano estacado," an incessant tremulous motion in the lower strata of the atmosphere, accompanied by a most singular and illusive mirage. This phenomenon, which so bitterly deluded the French army in Egypt, and has been observed in many other places, is here seen in perfection.

The very extraordinary refraction of the atmosphere upon these elevated plateaus, causes objects in the distance to be distorted into the most wild and fantastic forms, and often exaggerated to many times their true size. A raven, for instance, would present the appearance of a man walking erect; and an antelope often be mistaken for a horse or buffalo. In passing along over this thirsty and extended plain in a warm day, the eye of a stranger is suddenly gladdened by the appearance of a beautiful lake, with green and shady groves directly upon the opposite bank. His heart beats with joy at the prospect of speedily luxuriating in the cool and delicious element before him, and he urges his horse forward, thinking it very strange that he does not reach the oasis. At one time he imagines that he has made a sensible diminution in the distance, and goes on with renewed vigor and cheerfulness; then again he fancies that the object recedes before him, and he becomes discouraged and disheartened. And thus he rides for miles and miles, and still finds himself no nearer the goal than when first he saw it—when, perhaps, some sudden change in the atmosphere would dissipate the illusion, and disclose to him the fact that he had been following a mirage.

June 21.—On leaving our camp of last night, we crossed the creek and continued a south course for about five miles, when we rose upon the crest of a very elevated ridge which divides the waters of the north from those of the middle or Salt Fork; the valleys of both of which can be seen from this position. Descending upon the south side of the ridge, we encamped upon an affluent of the south fork, which runs rapidly through a narrow valley in an easterly course. The water is abundant, and free from salts.

The geological formation upon this side of the dividing ridge is different from that upon the north side, being here a soft, coarse, friable, conglomerated sandstone, enclosing a small drift of quartz, felspar, mica, and serpentine. The country in this vicinity is much broken and cut up with deep gorges and abrupt ridges, which are mostly impassable for wagons, and we have been obliged in consequence to travel a very circuitous route to-day, keeping the dividing ridges as much as possible, where we invariably find good ground for a road.

June 22.—In our course this morning, we struck one of the principal branches of the Salt Fork near its source, and followed it down upon the left bank to its confluence with the main stream. Below the junction the stream was fifty yards wide, but only about one-fourth of its bed covered with water. This branch of Red river, like the other, heads in the border of the "Llano estacado," and directly at the source is an elevated hill with abrupt vertical sides, terminating in a level summit; below this, upon the south bank, are two round mounds that can be seen for many miles.

We were much gratified in finding the water at the head of this branch, as in the north fork, sweet and wholesome. This settles the question that these branches of the river do not take their rise in salt plains, as has heretofore been very generally supposed. On the contrary, at their sources, which are in the eastern borders of the "Llano estacado," the water is as pure and wholesome as can be desired. And this character continues upon all the confluents until they enter the gypsum formation, when they become impregnated with salts, that impart a new character to the water, which continues to its junction with the Mississippi.

A solitary cotton-wood, with an occasional clump of willows, constitute the sylva of this portion of the river. The soil in the valley is an arenaceous red alluvium, and would be productive with the aid of artificial irrigation.

The bluffs bordering the valley are, at this place, about one hundred feet high, and composed of a deep red clay, overlaid with a stratum of drift; and this surmounted with a capping of calcareous sandstone from five to fifteen feet thick.

Upon the rocky bluffs bordering the river we found silicified wood in great quantities, strewed about over a distance of two miles. The petrifaction was most perfect, exhibiting all the fibres, knots, and bark, as plainly as in the native state, and was quite similar to the cotton-wood.

This evening we have another rain coming from the northwest, which will increase our chances for finding water in advance.

As it will be seen by a reference to the meteorological tables, our barometer has, in almost every instance, been a certain index to the weather from the commencement of the march. Sometimes, indeed, it has exhibited a most extraordinary depression of the mercury for two or three days previous to a storm; but in no instance has it failed to rain before the instrument would resume its usual range.

During the last three summers which I have spent upon the plains, as has been before observed, I have seen no rain of consequence from about the middle of May to the middle of August. And after passing west beyond the ninety-ninth degree of longitude, there has been but very little dew during the same period. The water in most of the streams was, at the same time, absorbed by the parched and porous soil over which it passed, and vegetation suffered much from the drought.

On the contrary, we have this season been favored with frequent and copious rains, and heavy dews. The streams have everywhere furnished a plentiful supply of good water, and the whole face of the prairies has been cheered with a rich and verdant vegetation. Near the place where we have pitched our tents this evening is an old Indian encampment, where John Bushman, our Delaware interpreter, has discovered that a battle has been fought within the past two months. The evidences of this are apparent from the fact that the remains of a large fire were found, upon which the victorious party had piled up and burned the lodges and effects of the vanquished. Pieces of the lodge-poles, and a quantity of fused glass beads, with small pieces of iron and other articles pertaining to their domestic economy, which had partially escaped the conflagration, were found scattered about the encampment. The number of lodge-fires indicated that the vanquished party was small.

The trail of a large party of Kioways, travelling to the north just before the last rain, has been seen to-day; and we are continually meeting with evidences of their having frequently resorted to this branch of the river. Their old camping-places and their trails are seen almost every day. They are probably at this time north of the Canadian, with the buffaloes; but are attracted to the waters of Red river in the autumn and winter, where the exuberant and rich grama grasses which everywhere abound in the river bottoms afford the finest pasturage to their numerous animals.

We have been gradually and regularly ascending in our progress westward, until now our approximate elevation above the sea, as indicated by the barometer, is two thousand seven hundred and two feet.

Our route to-day along the river valley has been populous with prairie dogs, their towns occupying almost the entire valley of the river. I was anxious to obtain a good specimen, and killed several of the largest I could find; but my rifle-ball mutilated them so much, that we did not think them worth preserving.

Our hunters brought in two deer and a turkey this evening, and their auxiliaries, the grayhounds, have added another deer to the list.

June 23.—This morning being dark, cloudy, and threatening rain, we did not leave camp until a late hour, when we continued our march down the left bank of the river for some four or five miles, directly at the base of the lofty escarpments of red clay and sandstone which terminate the valley upon the north side.

Soon after we started it commenced raining violently, and has continued incessantly throughout the day. It has raised the water in the river about twelve inches, so that now the entire bed is covered. In consequence of the rain we made an early encampment upon the south bank of the river.

The country upon each side of the river along where we have passed to-day has been much broken up into deep gorges and precipitous ridges, which are wholly impassable for wagons; and the features of the country adjoining have assumed a desert character. With the exception of a narrow strip of land forming the river-bottom, no arable soil can be seen, and no timber is found except a few stunted cottonwoods directly upon the river-banks. Several varieties of the wild sensitive plant, and especially the *Schrankia angustata*, are found everywhere throughout this section, and the atmosphere is redolent with the delightful perfume which is emitted from their blossoms.

Having traced this branch of the river to its source, and satisfied myself, from the portion that we have passed over, as to its general physical and topographical features, I have resolved to leave it at this point, and taking a southerly course, shall endeavor to make our way to the south branch of the river. I think the remainder of the time we have at our disposal can be more profitably occupied in exploring the country along the borders of that stream than in any other way.

We shall set out with a supply of water and wood sufficient for three days; and we hope, before that time expires, to find ourselves upon the waters of the south branch. Our animals that were poor when we left the settlements, are at this time in most excellent condition; and if we continue to find water and grass as abundant as we have done, we shall take them home in much better plight than they were at the commencement of our journey.

Thus far we have been most singularly fortunate in not losing even an animal by death or straying away; and, indeed, we have been much favored in every respect. The command have generally been in fine health and spirits, and with the exception of two cases of scurvy that originated before our departure from Fort Belknap, we have had no sickness worth mentioning.

June 24.—We were in motion at a very early hour this morning, and taking a southerly course directly at right-angles to the river, we soon became involved in a labyrinth of barren sand-hills, in which we travelled some fourteen miles before we emerged upon a high ridge, from which, in the distance, we could discern through the dim and murky atmosphere a very broad valley, through which we supposed the south branch to flow.

The bare and hot sand over which we had just passed was in strong contrast with the refreshing verdure of the valley before us. After travelling a few miles down the south slope of the divide, we encamped upon a small branch, where we found good water and grass, with a few cotton-wood trees, which furnished us with fuel.

The geological formation upon the bluffs bordering this stream is a friable red sandstone, overlaid with a stratum of coarse gypsum, with a subjacent stratum of bright red clay, interstratified with seams of gypsum. The soil since we left the sand-hills has been good, probably owing to the fertilizing properties of the gypsum.

June 25.—The atmosphere this morning was clear, cool, and bracing, with a north-northeasterly wind; the thermometer at 3 a. m. standing at 69°. The sky at sunrise was cloudless, and the sun shone brilliantly upon some elevated white bluffs which we could see in the distance, and supposed to be upon the border of the valley of the south fork of Red river.

At an early hour we resumed our march down the creek for about three miles, when we crossed another large stream with clear running water, and taking a circuitous course among the rough and broken hills bordering it, we made fifteen miles, encamping upon a branch where we found water standing in pools.

Our course to-day has led us through a formation of sulphate and carbonate of lime, which in some places appeared to be decomposed and covered the earth in a powdered state to the depth of three inches. Several fossil shells belonging to the cretaceous system were found to-day: they were much rounded by attrition, and probably have been transported here from a distance by water.

June 26.—We were in motion at the usual time this morning, and turning our course up the river over a very broken and elevated country, travelled ten miles, when we encamped upon a large branch of the south fork which enters from the north. It is fifty yards wide, with a sandy bed, and at this time contains but little water. The white escarpment of the Staked Plain has been in sight for the last two days in front and on the right of us. It seems to be very much elevated above the adjoining country, with almost vertical sides, covered with a scrubby growth of dwarf cedars, and from the summit the country spreads out into a perfectly level plain, or mesà, as far as the eye can penetrate.

The stream upon which we are encamped, like the other branches of Red river, takes its rise in the borders of this plain, and for several miles from its source there are numerous branches issuing from deep cañons, with perpendicular sides, which continue until they debouch into the more rolling country below, where the banks become low, and the bed broad and sandy.

The geological features of the country upon the head of this branch are characterized by a different formation from that upon the other branches we have seen, inasmuch as we here find the gypsum extending to the very sources, and the water having the peculiar taste imparted by that mineral throughout its entire course.

Our road during the whole day has passed through a continuous dog-town, (*Spermophilus ludovicianus*,) and we were often obliged to turn out of our course to avoid the little mounds around their burrows.

In passing along through these villages the little animals are seen in countless numbers sitting upright at the mouths of their domicils, presenting much the appearance of the stumps of small trees; and so incessant is the clatter of their barking, that it requires but little effort of the imagination to fancy oneself surrounded by the busy hum of a city.

The immense number of animals in some of these towns, or warrens, may be conjectured from the large space which they sometimes cover. The one at this place is about twenty-five miles in the direction through which we have passed it. Supposing its dimensions in other directions to be the same, it would embrace an area of six hundred and twenty-five square miles, or eight hundred and ninety-six thousand acres. Estimating the holes to be at the usual distances of about twenty yards apart, and each burrow occupied by a family of four or five dogs, I fancy that the aggregate population would be greater than any other city in the universe.

This interesting and gregarious little specimen of the mammalia of our country, which is found assembled in such vast communities, is indigenous to the most of our far western prairies, from Mexico to the northern limits of the United States, and has often been described by travellers who have been upon the plains. But as there are some facts in relation to their habits which I have never seen mentioned in any published account of them, I trust I shall be pardoned if I add a few remarks to what has already been said. In the selection of a site or position for their towns they appear to have a regard to their food, which is a species of short wiry grass, growing upon the elevated plains, where there is often no water near. I have sometimes seen their towns upon the elevated table-lands of New Mexico, where there was no water upon the surface of the ground for twenty miles, and where it did not seem probable that it could be obtained by excavating to the depth of a hundred feet. This has induced me to believe that they do not require that element without which most other animals perish in a short time.

As there are generally no rains or dews during the summer months upon the plains where these towns are found, and as the animals never wander far from home, I think I am warranted in coming to the conclusion that they require no water beyond that which the grass affords them. That they hybernate and pass the winter in a lethargic or torpid state is evident, from the fact that they lay up no sustenance for the winter, and that the grass around their holes dries up in the autumn, the earth freezes hard and renders it utterly impossible for them to procure food in the usual manner.

When the prairie dog first feels the approach of the sleeping season, (generally about the last days of October,) he closes all the passages to his dormitory to exclude the cold air, and betakes himself to his brumal slumber with the greatest possible care. He remains housed until the warm days of spring, when he removes the obstructions from his door and again appears above ground as frolicsome as ever.

I have been informed by the Indians that a short time before a cold storm in the autumn, all the prairie dogs may be seen industriously occupied with weeds and earth, closing the entrances to their burrows. They are sometimes, however, seen reopening them while the weather is still cold and stormy, but mild and pleasant weather is always certain to follow.

It appears, therefore, that instinct teaches the little quadrupeds when to expect good or bad weather, and to make their arrangements accordingly. A species of small owl is always found in the dog towns, sitting at the mouths of the holes when not occupied by the dogs; whether for the purpose of procuring food, or for some other object, I do not know. They do not, however, as some have asserted, burrow with the dogs; and when approached, instead of entering the holes, they invariably fly away. It has also been said that the rattle-snake is a constant companion of the dog; but this is a mistake, for I have sometimes passed for days through the towns without seeing one. They are, however, often seen in the holes in company with the dogs, and it has been supposed by some that they were welcome guests with the proprietors of the establishments; but we have satisfied ourselves that this is a domestic arrangement entirely at variance with the wishes of the dogs, as the snakes prey upon them, and must be considered as intruders. They are probably attracted to the burrows for the purpose of procuring food, as one snake which we killed was found to have swallowed a full-grown dog.

CHAPTER VI.

ARRIVE AT MAIN SOUTH FORK—PANTHER KILLED—BITTER WATER—INTENSE THIRST—HEAD SPRING—BEARS ABUNDANT—DEPARTURE DOWN THE RIVER.

June 27.—Making an early start this morning, we travelled down the river for five miles, when we crossed and resumed the south course over high rolling lands, much broken up on each side into numerous deep defiles and rugged cliffs, running towards the main river.

Directly in front of us lay the high table-lands of the "Llano estacado," towering up some eight hundred feet above the surrounding country, and bordered by precipitous escarpments capped with a stratum of white gypsum, which glistened in the sun like burnished silver. After travelling fourteen miles, we reached the valley of the principal branch of the river.

It was here nine hundred yards wide, flowing over a very sandy bed, with but little water in the channel, and is fortified upon each side by rugged hills and deep gullies, over which I think it will be impossible to take our train. The soil throughout this section is a light ferruginous clay, with no timber except a few hackberry and cotton-wood trees upon the banks of the streams. There is but little water either in the river or in the creeks, and in a dry season I doubt if there would be any found here.

Our route to-day has continued to lead us through dog towns, and it is probable that the fact of their being so abundant here has suggested the name which the Comanches have applied to this branch of Red river, of "Ke-che-a-qui-ho-no," or "Prairie-dog-town river."

We were so unfortunate yesterday as to lose an excellent bear-dog which a gentleman in Arkansas had taken great pains to procure for me. I regret this very much, as we are now coming into a country where we shall probably find these animals abundant, and it is difficult to hunt them without a good dog, trained for the purpose.

Our hunters killed two antelopes to-day. We have seen but few deer, however, and no turkeys, during the last week. We occasionally see the pinnated grouse and the quail; as also the meadow-lark, which I have found in all places wherever I have travelled.

June 28.—On leaving our encampment of last night, we took a southwesterly course for the eastern extremity of the white-capped bluffs which have been so long in sight, and which border the great plain of the "Llano estacado" upon the river valley.

After marching eight miles over a succession of very rugged hills and valleys, which rise as they recede from the river, we reached the base of these towering and majestic cliffs, which rise almost perpendicularly from the undulating swells of prairie at the base, to the height of eight hundred feet, and terminate at the summit in a plateau almost as level as the sea, which spreads out to the south and west like the steppes of Central Asia, in an apparently illimitable desert.

I supposed, from the appearance of the country at a distance, that I should be able to find a passage for the wagons along at the foot of these cliffs; but, upon a closer examination, find the ground between them and the river so much cut up by abrupt ridges and deep glens, that it is wholly impracticable to take our train any further up this branch of the river. We have sought for a passage by which we might take the train to the top of the bluffs, where, as they run nearly parallel to the course of the river, we might have continued on with the wagons; but after making a careful examination, we have abandoned the idea, not being able to discover a place where we could even take our horses up the steep sides of the precipice.

The geological formation of these bluffs is a red indurated clay, resting upon a red sandstone, overlaid with a soft, dark-gray sandstone, and the whole capped with a white calcareous sandstone, the strata resting horizontally, and receding in terraces from the base to the summit.

As Capt. McClellan and myself were passing to-day along under the bluffs, we saw in advance of us a herd of antelopes quietly feeding among some mezquite trees, when the idea occurred to me of attempting to call them with a deer-bleat, which one of the Delawares had made for me. I accordingly advanced several hundred yards to near the crest of a hill, from which I had a fair view of the animals, and, very deliberately seating myself upon the ground, screened from their observation by the tall grass around me, I took out my bleat and commenced exercising my powers in imitating the cry of the fawn. I soon succeeded in attracting their attention, and in a short time decoyed one of the unsuspicious animals within range of my rifle, which I raised to my shoulder, and, taking deliberate aim, was in the act of pulling trigger, when my attention was suddenly and most unexpectedly drawn aside by a rustling which I heard in the grass to my left. Casting my eyes in that direction, to my no small astonishment I saw a tremendous panther bounding at full speed directly towards me, and within the short distance of twenty steps. As may be imagined, I immediately abandoned the antelope, and, directing my rifle at the panther, sent a ball through his chest, which stretched him out upon the grass about ten yards from where I had taken my position. Impressed with the belief that I had accomplished a feat of rather more than ordinary importance in the sporting line, I placed my hand to my mouth, (" a la savage,") and gave several as loud shouts of exultation as my weak lungs would admit, partly for the purpose of giving vent to my feelings of triumph upon the occasion, and also to call the Captain, whom I had left some distance back with the horses. As he did not hear me I went back for him, and on returning to the spot where I had fired upon the panther, we discovered him upon his feet, making off. The Captain gave him another shot as he was running, and then closed in with his rifle clubbed, and it required several vigorous blows, laid on in quick succession, to give him his quietus.

The panther had probably heard the bleat, and was coming towards it with the pleasant anticipation of making his breakfast from a tender fawn; but, fortunately for me, I disappointed him. It occurred to me afterwards that it would not always be consistent with ones safety to use the deer-bleat in this wild country, unless we were perfectly certain we should have our wits about us in the event of a panther or large bear (which is often the case) taking it into his head to give credence to the counterfeit. This was a large specimen of the *Felis concolor*, or North American cougar, measuring eight feet from his nose to the end of the tail.

June 29.—As we were unable to proceed further up this branch of the river with the wagons, I concluded to leave the main body of the command under charge of Lieut. Updegraff, and, with Capt. McClellan and a small escort of ten men, to push on and endeavor to reach the head spring of this the principal branch of Red river.

Taking provisions for six days, packed upon mules, we went forward this morning over a constant succession of steep, rocky ridges, and deep ravines, in one of which we discovered a grotto in the gypsum rocks, which appeared to have been worn out by the continued action of water, leaving an arched passway, the sides of which were perfectly smooth and symmetrical, and composed of strata of three distinct bright colors of green, pink, and white, arranged in such peculiar order as to give it an appearance of singular beauty. On our arrival here the men were much exhausted by rapid marching over the rough ground, and were exceedingly thirsty. Fortunately, we found near the mouth of the grotto a spring of very cold water bursting out of the rock; and although it had the peculiar taste of the gypsum, yet they drank large quantities without suffering from it.* Our animals and men being much jaded from travelling over this rough and forbidding country, we turned down towards the river after a short halt at the grotto, and on reaching it found the water still very bitter and unpalatable. As the day was very warm, (the thermometer standing, at 12 o'clock m., at 104° Fahrenheit in the shade) with no air stirring, the reflection of the sun's rays from the white sand in the bed of the river made it exceedingly oppressive.

At sundown we bivouacked near a small pool of muddy water, a little better than that in the river, but still very unpalatable. In despite of this, as we were suffering much from the intense thirst caused by the heat of the day, and from drinking the nauseating water we had met with upon the march, we indulged freely; but instead of allaying thirst it only served to increase it.

The country over which we have passed to-day, upon both sides of the river, has been cut up by numerous deep gorges extending from the chain of mural escarpments that terminate the "Llano estacado" to the river, and in many of these are small streams of water which issue from springs in the rocky sides of the gorges. We have met with no

---

* As this spring issued directly from the pure gypsum rock, I procured a specimen of the water, which has been analyzed under the direction of Professor W. S. Clark, in the laboratory of Amherst College, and may, I think, be regarded as containing those ingredients which communicate that peculiar disagreeable taste to all the water of this country that flows over a gypsum formation. The analysis resulted as follows:

Water, in fluid ounces	4
" in fluid grammes	127.500
Hydrosulphuric acid present	.011
Chlorine	.014
Lime	.090
Sulphuric acid	.227
Soda and magnesia, about	.130

These elements, united in the form of salts, would give the following results:

Weight of sulphate of lime	.219
" " " magnesia	.088 (?)
" " " soda	.073 (?)
" " chloride of sodium	.023
" " hydrosulphuric acid	.011
Weight of the whole	.414
Per-centage of matter in solution	0.82

trees except a species of red cedar, *Juniperus Virginiana*, and a few lonely cotton-woods.

The soil is sandy upon the ridges, with blue and red clay in the valleys, and gypsum rocks predominate throughout the formation. The high bluffs to the south of us have gradually approached the river until, near our encampment, they are only about two hundred yards distant.

June 30.—At daylight this morning we were in the saddle, and, taking the bed of the river, set out at a brisk pace, hoping to find some good water during the day. Our course was very circuitous from being obliged to follow the windings made by the numerous detours in the river. The lofty escarpments which bounded the valley upon each side, rose precipitously from the banks of the river to the enormous height of from five to eight hundred feet; and in many places there was not room for a man to pass between the foot of the acclivities and the river. It was altogether impossible to travel upon either side of the river, so much broken and cut up was the ground, and the only place where a passage for a horse can be found is directly along the defile of the river bed. We found frequent small rivulets flowing into the river through the deep glens upon each side; but, most unfortunately for us, the water in them all was acid and nauseating. We made our noon halt at one of these streams, after travelling fifteen miles over the burning sands of the river bed.

At this time we had become so much affected by the frequent and irresistible use of the water, that most of us experienced a constant burning pain in the stomach, attended with loss of appetite, and the most vehement and feverish thirst. We endeavored to disguise the taste of the water by making coffee with it, but it retained the same disagreeable properties in that form that it had in the natural state.

At four in the evening we again pushed forward up the river, praying most devoutly that we might reach the termination of the gypsum formation before night, and that the river, which was still of very considerable magnitude, would branch out and soon come to a termination.

Four miles from our halting-place we passed a large affluent coming in from the north, above which there was a very perceptible diminution in the main stream; and in going a few miles further, we passed several more, causing a still greater contraction in its dimensions. All these affluents were similar in character to the parent stream, bordered with lofty and precipitous bluffs, with gypsum veins running through them similar to those upon the main river.

Towards evening we arrived at a point where the river divided into two forks, of about equal dimensions. We followed the left, which appeared somewhat the largest, and here found the bluffs receding several hundred yards from the banks upon each side, leaving a very beautiful and quiet little nook, wholly unlike the stern grandeur of the rugged defile through which we had been passing. This glen was covered with a rich carpet of verdure, and embowered with the foliage of the graceful china and aspen, and its rural and witching loveliness gladdened our hearts and refreshed our eyes, long fatigued with gazing upon frowning crags and deep, shady ravines.

After travelling twenty-five miles we encamped upon the main river, which had now become reduced to one hundred feet in width, and flowed rapidly over a sandy bed.

Although we were suffering most acutely from the effects of the nauseating and repulsive water in the river, yet we were still under the painful necessity of using it. Several of the men had been taken with violent cramps in the stomach and vomiting, yet they did not murmur; on the contrary, they were cheerful, and indulged in frequent jokes at the expense of those who were sick. The principal topic of conversation with them seemed to be a discussion of the relative merits of the different kinds of fancy iced drinks which could be procured in the cities, and the prices that could be obtained for some of them if they were within reach of our party. Indeed it seems to me that we were not entirely exempt from the agitation of a similar subject; and from the drift of the argument, I have no doubt that a moderate quantity of Croton water, cooled with Boston ice, would have met with as ready a sale in our little mess, as in almost any market that could have been found. If I mistake not, one of the gentlemen offered as high as two thousand dollars for a single bucket of the pure element; but this was one of those few instances in which money was not sufficiently potent to attain the object desired.

We laid ourselves down upon our blankets and endeavored to obliterate the sensation of thirst in the embraces of Morpheus; but so far as I was concerned, my slumbers were continually disturbed by dreams, in which I fancied myself swallowing huge draughts of ice-water.

July 1.—We saddled up at a very early hour this morning, and proceeded on up the river for several miles, when we found a large affluent putting in from the north; and after travelling a few miles further we passed many more small tributaries, which caused the main stream to contract into the narrow channel of only twenty feet; and its bed, which from its confluence with the Mississippi to this place (with the exception of a ridge of rocks which crosses it near Jonesborough, in Texas) had been sand, suddenly changed to rock, with the water, which before had been turbid, flowing clear and rapidly over it; and, much to our delight, it was entirely free from salts. This was certainly an unlooked-for luxury, as we had everywhere before this found it exceedingly unpalatable. As I before observed, the effect of this water upon us had been to produce sickness at the stomach, attended with loss of appetite, and a most raging and feverish thirst, which constantly impelled us to drink it, although it had a contrary effect upon us from what we desired, increasing rather than allaying thirst.

After undergoing the most intense sufferings from drinking this nauseating fluid, we indulged freely in the pure and delicious element as we ascended along the narrow dell through which the stream found its way. And following up for two miles the tortuous course of the gorge, we reached a point where it became so much obstructed with huge piles of rock, that we were obliged to leave our animals and clamber up the remainder of the distance on foot.

The gigantic escarpments of sandstone, rising to the giddy height of eight hundred feet upon each side, gradually closed in until they were only a few yards apart and finally united over head, leaving a long, narrow corridor beneath, at the base of which the head spring of the principal or main branch of Red river takes its rise. This spring bursts out from its cavernous reservoir, and, leaping down over the huge masses of rock below, here commences its long journey to unite with other tributaries in making the Mississippi the noblest river in the universe. Directly at the spring we found three small cotton-wood trees, one of which was blazed, and the fact of our having visited the place, with the date, marked upon it.

On beholding this minute rivulet as it wends its tortuous course down the steep descent of the cañon, it is difficult to realize that it forms the germ of one of the largest and most important rivers in America; floating steamers upon its bosom for nearly two thousand miles, and depositing an alluvion along its borders which renders its valley unsurpassed for fertility.

We took many copious draughts of the cool and refreshing water in the spring, and thereby considered ourselves, with the pleasure we received from the beautiful and majestic scenery around us, amply remunerated for all our fatigue and privations. The magnificence of the views that presented themselves to our eyes as we approached the head of the river, exceeded anything I had ever beheld. It is impossible for me to describe the sensations that came over me, and the exquisite pleasure I experienced, as I gazed upon these grand and novel pictures.

The stupendous escarpments of solid rock, rising precipitously from the bed of the river to such a height as, for a great portion of the day, to exclude the rays of the sun, were worn away, by the lapse of time and the action of the water and the weather, into the most fantastic forms, that required but little effort of the imagination to convert into works of art, and all united in forming one of the grandest and most picturesque scenes that can be imagined. We all, with one accord, stopped and gazed with wonder and admiration upon a panorama which was now for the first time exhibited to the eyes of civilized man. Occasionally might be seen a good representation of the towering walls of a castle of the feudal ages, with its giddy battlements pierced with loopholes, and its projecting watch-towers standing out in bold relief upon the azure ground of the pure and transparent sky above. In other places our fancy would metamorphose the escarpments into a bastion front, as perfectly modelled and constructed as if it had been a production of the genius of Vauban, with redoubts and salient angles all arranged in due order. Then, again, we would see a colossal specimen of sculpture representing the human figure, with all the features of the face, which, standing upon its lofty pedestal, overlooks the valley, and seems to have been designed and executed by the Almighty artist as the presiding genius of these dismal solitudes.

All was here crude nature, as it sprang into existence at the fiat of the Almighty architect of the universe, still preserving its primeval type, its unreclaimed sublimity and wildness; and it forcibly inspired me with that veneration which is justly due to the high antiquity of nature's handiworks, and which seems to increase as we consider the solemn and important lesson that is taught us in reflecting upon their continued permanence when contrasted with our own fleeting and momentary existence.

On climbing up to the summit of the escarpment over the head of the spring, we found ourselves upon the level plain of the "Llano estacado," which spreads out from here in one uninterrupted desert, to the base of the mountains east of the Rio Grande. The geographical position of this point, as determined by courses and distances from the place where we left the wagons, is in latitude 34° 42' north, and longitude 103° 7' 11" west; and its approximate elevation above the sea, as determined by frequent and careful barometric observations, is 2,450 feet.

The geological formation is different here from what it is below, inasmuch as we find no gypsum; and the moment we passed this mineral, (which was only about two miles before we reached the head of the river,) the water became at once sweet and good.

We have seen numerous bear tracks within the past two days; and occasionally the animals themselves, two of which we killed. Several that we saw, however, escaped; and we had frequent occasion to regret the loss of our bear-dog, as we might have killed many more with his assistance.

John Bull, who still continued to ride the same fractious horse which he had in the buffalo hunt, made a brush with a large bear to-day, but did not succeed in getting alongside of him, as the horse became perfectly mad and unmanageable the moment he got sight of the bear. This is often the case; and there are but few horses that can be made to approach one of these animals.

Several anecdotes, which were related to me by our guide, concerning the habits of the black bear, would seem to entitle him to a higher position in the scale of animal instinct and sagacity than that of almost any other quadruped. For instance, he says that before making his bed to lie down, the animal invariably goes several hundred yards with the wind, at a distance from his first track. Should an enemy now come upon his track, he must approach him with the wind; and with the bear's keen sense of smell, he is almost certain to be made aware of his presence, and has time to escape before he is himself seen.

He also states that when pursued, the bear sometimes takes refuge in caves in the earth or rocks, where the hunter often endeavors, by making a smoke at the entrance, to force him out; but it not unfrequently happens, that instead of coming out when the smoke becomes too oppressive, he very deliberately advances to the fire, and with his fore feet beats upon it until it is extinguished, then retreats into the cave. This he assured me he had seen often. Although these statements would seem to endow bruin with something more than mere animal instinct, and evince a conception of the connexion between cause and effect, yet another anecdote which was related to me would go to prove this curious quadruped one of the most stupid fellows in the brute creation.

My informant says, that when the bear cannot be driven out of the cave by smoke, it sometimes becomes necessary for the hunter to take his rifle, and with a torch to enter the cavern in search of him. One would suppose this a very hazardous undertaking, and that the animal would soon eject the presumptuous intruder; but, on the contrary, as soon as he sees the light approaching, he sits upright on his haunches, and with his fore paws covers his face and eyes, and remains in this position until the light is removed. Thus the hunter is enabled to approach as close as he desires without danger, and taking deadly aim with his faithful rifle, poor bruin is slain. These facts have been stated to me by three different Indians, in whose veracity I have much confidence, and I have no doubt are strictly true. The black bear is generally harmless unless wounded, or when accompanied by its young, when I have known one of them to pursue a man on horseback several hundred yards in the most furious mood, snapping continually at the legs of the horse.

July 3.—We reached camp to-day from the head of the river, having returned over the same route that we ascended, and found all anxiously awaiting us. From this point to the head of the river is sixty-five miles, and for about sixty miles of this distance the river runs through a deep defile, the escarpments of which rise from five to eight hundred feet upon each side, and in many places they approach so near the water's edge that there is not room for a man to pass, and it is often necessary to travel for several miles in the bed of the river before a place is found where a horse can clamber up the precipitous sides of the chasm.

I could not determine in my own mind whether this remarkable defile had been formed, after a long lapse of time, by the continued action of the current, or had been produced by some great convulsion of nature: perhaps both causes have contributed to its formation, some convulsive operation having first given birth to an extensive fissure, and the ceaseless action of the stream having afterwards reduced it to its present condition.

A gentleman who is travelling with us, and who was attached as a captain to Col. McLeod's expedition to Santa Fe, so graphically described by Mr. Kendall, recognised a point, near the head of the river, where his command passed. He is of the opinion that the river which they ascended, and supposed at the time to be the principal branch of Red river, must have been the Big Witchita, and they probably passed entirely to the south of the main branch of the river. The fact that they were for a long time upon the plains of the "Llano estacado" would go to confirm this supposition, as anywhere to the north of this stream they would not have encountered much of it.

July 4.—This morning at an early hour we turned our faces towards home, and travelled about five miles down the right bank of the river, when we discovered that the country in advance upon that side was so much broken into deep gullies and abrupt ridges that it would be impracticable to get our wagons over them. We therefore crossed to the north side of the river, where we found a most excellent road over smooth prairie. At our present position we have a pond of excellent water, with an abundance of hackberry and cotton-wood for fuel. On approaching the pond, Capt. McClellan and myself, who were in advance of the command, espied a huge panther very leisurely walking away in an opposite direction; and as, in hunter's parlance, we "had the wind of him," it enabled us to ride sufficiently near to give him a shot before he discovered us. It took effect and caused him to make a tremendous leap into the air, and, running a short distance, he fell dead. We have also killed four deer to-day, which supplies us with an abundance of fresh meat. Some of the bucks are now very fat, and the venison is superior to any I have ever eaten.

The pond of water at our camp is a very peculiar and strange freak of nature. It is almost round, two hundred and fifty feet in diameter, with the water thirty feet deep, and perfectly transparent and sweet The surface of the water in this basin is about twenty feet below the banks, and the sides of the depression nearly perpendicular. The country for two or three miles around, in all directions, rises to the height of from one to two hundred feet. As this pond seems to be supplied by springs, and has no visible outlet, it occurred to me that there might possibly be a subterraneous communication which carried off the surplus water and the earth from the depression of the basin.

July 5.—We were in motion this morning at 2 o'clock, keeping down the left bank of the river, in an easterly course over a firm and smooth road for sixteen miles, when we found ourselves upon a small running creek, the water of which was strongly charged with salts; but as we had filled our casks at the pond, we did not suffer.

We are encamped near a conical-shaped mound, flat upon the top, and are about three miles from the main river.

We find much more mezquite timber upon this branch of the river than upon the other. Indeed, I have never seen much of this wood above the thirty-sixth degree of north latitude; but south of this it appears to increase in quantity and size as far as the 28th degree. Upon the Canadian river I have observed a few small bushes; but the climate in that latitude appears too cold for it to flourish well.

The soil here is sandy, with but little water, and that for the most part of a quality unfit for use. The grama and mezquite grasses are abundant. Our route for the last fifty miles has carried us through an almost continuous dog-town, but as yet we have not been able to secure a live specimen. The latitude at this point is 34° 8' 30".

July 6.—Our wagons were packed, and we were *en route* before 3 o'clock this morning, but were obliged to deviate from our course very considerably to pass around some deep ravines that extended back to near the crest of the ridge, dividing the middle from the south fork. In this route we traversed a very smooth and elevated rolling prairie, from which we frequently obtained views of the valleys of both branches of the river.

The grama grass, which appears to flourish in this section, is now in process of heading, and will soon be matured.* This most excellent forage for animals does not ripen until quite late in the season, and remains green during most of the winter. I have observed it growing in about the same latitudes as the mezquite trees; but it is most abundant in New Mexico, where it is the predominating grass of the country.

As I was riding at a distance from the train to-day, I saw three Indians, but they immediately passed out of view in a ravine, and were not observed again.

We are encamped this evening upon a very clear and rapid brook; but the water, unfortunately, has the characteristic taste of the gypsum.

There is capital grass upon the creek, and large cotton-wood and hackberry, with a few mulberry trees, which, being the first we have seen for several weeks, has suggested a name for the stream—"*Mulberry Creek.*"

July 7.—We left camp at 2 o'clock this morning, and continued on for three miles over the same description of country as that we passed yesterday, when we arrived at a swift-running creek, twenty-five feet wide and eight inches deep, of clear, cold water; but, as usual, upon tasting it, found it unpalatable. After passing this creek our course was nearly parallel to the river, and from four to twelve miles distant.

The gypsum formation characterizes this section, and has continued from near the head of the river to this place; but as it imparts to the water such disagreeable qualities, we earnestly desire to see no more of it.

** Two varieties of grama grass-seed (*Chondrosium foeneum* and *Atheropogon oligostachyum*) were collected and disposed of in the manner mentioned in the following letter:*

U. S. PATENT OFFICE,
November 12, 1852.

SIR: The two packages of grama grass-seed from near the sources of Red river, forwarded by you to this office, have been received, and you are requested to accept the thanks of the office for the same. They have already been distributed, in conformity with your suggestion, to gentlemen in the States of Virginia, North Carolina, South Carolina, Georgia, Alabama, and Louisiana.

Very respectfully, your obedient servant,

S. H. HODGES.

Capt. R. B. MARCY, *New York*.

One of our Delawares killed a very large wild cat (*Lyncus rufus*) to-day, the skin of which we have preserved.

Our collection of reptiles increases very rapidly, and we now have upwards of a hundred specimens, many of them very beautiful and interesting. Our herbarium is also enlarging daily, and we already have a large collection.

CHAPTER VII.

ANTELOPE AND DEER—WITCHITA MOUNTAINS IN SIGHT—REACH BUFFALO CREEK—VALLEY OF OTTER CREEK—SALUBRITY OF CLIMATE—DEER BLEAT—HORSE-FLIES—SCURVY—WITCHITA MOUNTAINS—PASS THROUGH THE MOUNTAINS—BUFFALO SEEN.

July 8.—Our train was in motion again at 2 o'clock this morning, and our road led us over very elevated table-lands, near the dividing ridge of the two branches of the river, where the country is totally destitute of wood or water, and altogether devoid of interest until reaching this place, where we find a few small ponds of wretched water and a clump of trees.

In addition to four deer and two antelopes that have been killed by our party to-day, our grayhounds have contributed another deer to our larder.

We have had several good opportunities since we have been upon the plains of witnessing the relative speed of the different animals found here, and our observations have confirmed the opinion I have before advanced. For example, the grayhounds have upon several different occasions run down and captured the deer and the prairie rabbits, which are also considered very fleet; but although they have had very many races with the antelope under favorable circumstances, yet they have never in one instance been able to overtake them; on the contrary, the longer the chase has continued, the greater has been the distance between them. The *Cervus Virginianus* (our red deer) has generally been considered the fleetest animal upon the continent after the horse, but the *Antilocapra Americana*, or prong-horned antelope of the plains, is very much swifter.

One of our hunters, who has been in advance of our camp, says he obtained a distant view of the Witchita mountains, and that he has also discovered several telegraphic smokes in a northeasterly direction.

July 9.—Getting under way at 2 o'clock this morning, we journeyed over the elevated prairie in a northeast course towards the dividing ridge, and on coming upon the crest of this elevation, some of the most lofty peaks at the western extremity of the Witchita chain of mountains showed themselves in the distance, like smoky clouds against the background of the murky sky near the horizon. Crossing over the ridge, we made for the head of a creek, where we expected to find good water, but upon reaching it we found the gypsum rocks, and, as usual, the water exceedingly bitter, and wholly unfit for use. After travelling down this creek for four miles, we encamped at a small pond, containing a liquid which we were obliged to make use of, but it had more the appearance of the drainings from a stable-yard than water.

We find more timber upon the borders of this stream than we have seen since leaving Sweet-water creek; it consists of china, hackberry, cotton-wood, and mulberry. The grass is luxuriant, and the vegetation of the valley has a smiling and verdant aspect, that marks the fertility of the soil.

Four deer have been killed to-day—two of which I was so fortunate as to add to my list: one was also caught by the grayhounds. They have afforded us much and rare sport by frequent chases, of which the smooth prairie has afforded us a good view.

It is a most beautiful spectacle to mark the slender and graceful figures of the hounds as they strain every muscle to its utmost tension in their eager and rapid pursuit of the panic-stricken deer. It is a contest between two of the fleetest and most graceful and beautiful quadrupeds in existence: the one has his life at stake, and the other is animated by all that eager enthusiasm which is characteristic of a thorough-breed animal. They both put forth all the energies with which the Author of their being has endowed them, and seem to fly over the wavy undulations of the plains. Now they are upon the summit of one of these swells, and the startled animal has disappeared in an adjoining ravine, and for a moment the hounds are at fault; but soon they espy him panting up the opposite acclivity, when they are off again like the wind, in hot pursuit, and, rapidly closing upon their devoted victim, they are soon engaged in the death struggle. This sport is most intensely exciting, and he who would not become interested in it would hardly be entitled to claim consanguinity with the great family of Nimrod.

The result of our observations for latitude at this position is $34° 8' 11''$.

July 10.—As the country over which we had to pass this morning was intersected by numerous abrupt ravines, we were unable to leave camp until daylight.

Our course led us over a high ridge, in an easterly direction for several miles, when we arrived upon the banks of a deep and rapid affluent of the main river, along which we travelled for two miles, encamping near a spring of cold, but brackish water.

We have seen Indian-tracks to-day, made about three days since, and are much astonished that they have not paid us a visit, as some of the different parties we have passed must have seen our trail.

The Witchita mountains have been in sight to the left all day, and our present position is very nearly opposite the western extremity of the chain. The variation of the magnetic needle at this point is 10° 45' 30" east.

July 11.—Striking our tents at an early hour this morning, we continued down the valley of the creek for ten miles, when we turned to the north and followed for several miles a ridge dividing this from another stream, upon which we are encamped.

The face of the country over which we are now journeying is totally without interest, being arid, sterile, and flat, and presenting no object upon which the eye can rest with pleasure.

The stream at this place is thirty yards wide, two feet deep, with a swift current, and the water brackish. Since we left the head of the Ke-che-a-qui-ho-no, we have found but three places upon the route where the water has been entirely free from salts, and at these places, with one exception, it has been insipid, stagnant, and muddy; yet our animals drink it and appear fond of it. As yet, we have lost none of our stock by death or straying. Our oxen, although they have performed more labor than the mules, are in much better condition; indeed, they have been constantly improving, while the others have become somewhat poor and jaded. This goes to confirm me in an opinion I had previously formed as to the comparative powers of endurance of the two different kinds of cattle for long journeys upon the plains. I have now no hesitation in expressing a decided opinion in favor of the oxen.

July 12.—As we anticipated a long march, réveille was sounded at 1 o'clock this morning, and we were *en route* at 2. Taking a course north of east towards a mountain which we recognised as being upon Beaver creek, we reached the confluence of this stream with Red river at 9 o'clock, and crossing a short distance above the junction, encamped in a bend of the creek, where, to the supreme satisfaction of every one in the command, we once more found good running-water, and after being for so long a time deprived of it we enjoyed it exceedingly.

When drinking the bad water upon the plains it has often occurred to me that we do not sufficiently appreciate the luxury of good water in those more favored parts of our country, where it everywhere abounds in the greatest profusion. The suffering produced by the absence of good water in a journey on the plains during the heat of the summer months is known only to those who have experienced it. As we have now passed the gypsum range of country, we do not anticipate any more difficulty in finding good water.

We shall remain at this place to-morrow, and on the day following propose to ascend Otter creek to the mountains, and passing down through the chain, shall make a careful and thorough examination of the geological character of the formation, and any other objects of interest that may present themselves in our route.

Red river, above the mouth of Otter creek, which was at a stage above fording when we passed up, is now only two feet deep, and flows at the rate of about three miles per hour.

Fresh buffalo-tracks have been seen to-day, and six deer and one turkey brought in by the hunters.

July 13.—This morning, for the first time in several weeks, we have had a rain, which has refreshed and revivified the whole face of the country. Previous to this the ground had become so much parched from the lack of moisture, that vegetation was suffering considerably. The herbage in the valley of the creek appears to have felt the drought more than upon the elevated prairies; here it has put on a yellow tinge, and a perfume is emitted from it similar to that of fresh hay, while upon the more elevated plains it still retains its deep green attire. Nine deer have been killed to-day, and I again marked two upon my list.

July 14.—Captain McClellan and myself started out this morning to make an examination of the country along the upper portion of the valley of the creek, while the command crossed and encamped about four miles above our position of last night.

There is much more woodland towards the sources of the stream than I had supposed. Black walnut, pecan, hackberry, elm, and cottonwood, are among the varieties of timber found here; the mezquite is also abundant near the mountains.

Many of the trees in the bottom are straight and of sufficient dimensions to make good building material, and there is an ample supply for the farmer's purposes. The soil in the valley is for the most part a dark, rich alluvium, sustaining a dense carpet of herbage, and I have no doubt would yield abundant crops of grain.

The stream extends in two principal branches back to the mountains, where they receive numerous small tributary rivulets flowing from springs. The course of the principal branch is northeast and southwest, and is about twenty miles in length. The mountains here appear to be in groups or clusters of detached peaks of a conical form, indicating a volcanic origin, with smooth, level glades intervening; and rising, as they do, perfectly isolated from all surrounding eminences upon the plateau of the great prairies, their rugged and precipitous granite sides almost denuded of vegetation, they present a very peculiar and imposing feature in the topographical aspect of the country. From the fact that the ground occupying the space between the mountains is a level, smooth surface, and exhibits no evidence of upheaval or distortion, may it not with propriety be inferred that the deposition here is of an origin subsequent to that of the upheaval of the mountains?

July 15.—We were in motion at 2 o'clock this morning, and taking a northeast course towards the base of the mountain chain, passed through mezquite groves, intersected with several brooks of pure water flowing into the south branch of Cache creek, upon one of which we are encamped.

We find the soil good at all places near the mountains, and the country well wooded and watered. The grass, consisting of several varieties of the grama, is of a superior quality and grows luxuriantly. The climate is salubrious, and the almost constant, cool, and bracing breezes of the summer months, with the entire absence of anything like marshes or stagnant water, remove all sources of noxious malaria, with its attendant evils of autumnal fevers.

I was so fortunate as to kill a very large and fat buck to-day, which adds much relish to the good cheer of our evening meal. Three others having been brought in by the hunters, our larder is at present well stocked with meat. Indeed, there has been but a small portion of the time since we have been out that our excellent hunters have not supplied the entire command with an abundance of fresh meat. Although we have beef-cattle in the train, we have as yet had no occasion to make use of one of them.

One of the Delawares has seen fresh buffalo-tracks to-day going to the southeast, and we still cherish the hope that we may yet encounter them.

John Bushman, our interpreter, was much surprised to-day, on calling a doe towards him with a deer-bleat, to see a small fawn following after its mother; but imagine his astonishment, when immediately behind the fawn came a huge panther bounding rapidly towards him, and in a twinkling he fastened his claws in the vitals of his victim. He, however, in this instance, caught a tartar, and paid dearly for his temerity, as John, with a spirit of indignation that would have done credit to the better feelings of any man, raised his rifle, and, instead of killing the deer, which was entirely at his mercy, planted the contents in the side of the panther.

The method of hunting deer by the use of the bleat is practised extensively by the Delawares in this country, and with great success.

They make the bleat somewhat similar to the first joint of a clarionet, with a brass reed scraped very thin, and applied in the same manner as upon the clarionet, and so regulate and adjust the instrument by experiment as to imitate almost precisely the cry of the young fawn. They use them during the months of June and July, before the does have weaned their young. Riding along near a copse of trees or brush where they suppose the deer to be lying, they sound their bleats, which can be heard for half a mile; and as the doe never remains near her fawn any longer than is necessary to give it food, (when she retires to an adjoining thicket and makes her bed alone,) she immediately takes alarm at what she conceives to be a cry of distress from her helpless offspring, and, in the intensity of her maternal affection, she rushes at full speed in the direction of the cry, and frequently comes within a few yards of the hunter, who stands ready to give her a death-wound. This is an unsportsman-like way of hunting deer, and only admissible when provisions are scarce.

The bear, the wolf, and panther often come at the call of the bleat, supposing they are to feast upon the tender flesh of the fawn. It might be supposed that in a country where there are so many carnivorous animals, the greater portion of the deer would be killed by them while young; but nature, in the wisdom of its arrangements, has provided the helpless little quadruped with a means of security against their attacks, which is truly wonderful. It is a well-known fact among hunters that the deer deposite a much stronger scent upon their tracks than any other animal, inasmuch as a dog can without difficulty follow them long after they have passed at a distance of many yards from the track. Notwithstanding this, the fawns, until they are sufficiently grown to be able to make good running, give out no scent whatever upon their tracks, and a dog of the best nose cannot follow them except by sight. I have often seen the experiment made, and am perfectly satisfied that such is the case; this, therefore, must in a great measure protect them from the attacks of the wild animals of the country.

July 16.—Our réveille sounded at two, and we were *en route* at 3 o'clock this morning. Continuing a northeast course for four miles, we crossed a fine stream of clear water issuing from the mountains and running into the south branch of Cache creek; after travelling three miles further we passed another, and made our encampment upon a third: all of these were of about equal magnitude, and similar in character. They take their rise from springs among the granite mountains, and flow over the detritus and sand at the base; are about twenty feet wide, with the water clear and rapid. The banks are abrupt, about ten feet high, and composed of white clay and sandstone. Upon each of these branches there are large bodies of post-oak timber, much of which would serve as building-material, and near the bank of the creek we observed black-walnut.

Within a distance of six miles around our camp, I should estimate the amount of wood-land at eight thousand acres. The grass is of the very best quality, and the soil cannot be surpassed for fertility.

We are, at this place, directly at the base of one of the most lofty and rugged mountains of the range. Its bare and naked sides are almost destitute of anything in the shape of a tree or plant, and it is only here and there that a small patch of green can be discerned. Huge masses of flesh-colored granite, standing out in jagged crags upon the lofty acclivities, everywhere present themselves to the eye, and the scenery is most picturesque, grand, and imposing.

We have for a few days past been much annoyed with a species of large, black horse-fly, which attacks the animals most savagely, and leaves his red mark wherever he touches them. These, with a species of small black gnat, are the only insects that we have been troubled with.

The two men who for several weeks have been suffering from the scurvy are no better, and I am fearful, if we do not find the wild onion soon, that they will be in a bad state.

I have caused all the men of the command to use freely what few anti-scorbutics we were enabled to procure from the subsistence department, as also all the wild vegetables that could be obtained upon the march; but these do not seem sufficient to fend off the disease, when men have for a long time been confined exclusively to animal diet and constantly subjected to other causes that predispose the system to the disease.

The soldiers are by no means anxious to make use of the anti-scorbutics from the commissary department, as they are obliged to pay for them by submitting to a deduction in the amount of their ration, which, at most, is a very small allowance for men who are marching or laboring hard. This fact is so well established, that when citizen teamsters are employed in the quartermaster's department, it is either necessary to give them an allowance of fifty per cent. more in the amount of provisions than the soldier gets, or an addition to his pay to enable him to purchase an equivalent. Doctor Shumard has made use of all the remedies in his possession in the cases of scurvy that have been under his treatment, but he is of opinion that they avail but little in the absence of vegetable diet. Our men have discovered some green grapes to-day, which I hope may relieve the sick men. Several gentlemen of the party ascended the mountain near our camp this evening, and obtained a fine view of the adjoining country. They discovered that there were three distinct ranges running from northeast to southwest; at this place they appear to be united in one chain, and there seems to be no pass practicable for wagons in this vicinity.

July 17.—Moving out from camp at half-past three this morning, we journeyed along the southeastern base of the mountains, passing several spring-brooks of cold, delicious water, flowing from the deep gorges of the mountain, over the masses of loose rock at the bases, into the valley below. These brooks are perennial, and this being the dry season, they are probably now at their lowest stage, yet there is a sufficiency of water for all purposes of farmers and for milling.

The soil continues of an excellent quality, and sustains a heavy vegetation. In addition to the advantages of rich soil, good timber, and water, which everywhere abound near the mountains thus far upon our route, may be added that of the great salubrity of the climate.

The atmosphere in these elevated regions is cool, elastic, and bracing, and the breezes which sweep across the prairie temper the heat of the sun, and render it, even in midsummer, cool and comfortable.

The different branches of Cache creek drain a large extent of country, which might be made available for agricultural purposes, and would be sufficient to sustain a large population.

The particular district embracing the Witchita mountains has for many years been occupied and (with much justice, it seems to me) claimed by the Witchita Indians, who have a tradition that their original progenitor issued from the rocks of these mountains, and that the Great Spirit gave him and his posterity the country in the vicinity for a heritage, and here they continued to live and plant corn for a long time.

Notwithstanding this claim of the Witchitas, which the right of occupancy and possession has guarantied to them, yet the whole of this beautiful country, as far as the 100th degree of west longitude, is included in the grant made by the United States to the Choctaws, who thereby possess the greater part of the lands upon Upper Red river that are really valuable.

The Witchitas are an insignificant tribe in point of numbers, not having more than about five hundred souls in the nation, and are not, of course, prepared to substantiate or enforce their title to this country; and, indeed, I very much doubt if they have any claims upon the consideration or generosity of our government, being the most notorious and inveterate horse-thieves upon the borders, as the early frontier settlers of Texas can vouch for; and they are only held in restraint now by fear of the troops near them. They have always been extremely jealous of the motives of the white people who have wished to penetrate to the interior of their country, and have, upon several occasions, driven off parties who have attempted to examine the country about the Witchita mountains.

We are encamped this evening upon a swift-running brook, near a very cold spring of pure water, which affords a delightful contrast to the water we have met with upon the Ke-che-a-qui-ho'-no. Following up the large brook into which the spring empties, I found its source in a most lovely valley, about two miles above our encampment.

This valley, which is enclosed on three sides by lofty and rugged mountains, is mostly covered with a heavy growth of timber of a very superior quality. The trees, which are oak, are large, straight, and tall, and are the best suited to the carpenter's purposes of any I have ever seen west of the "Cross-Timbers." The soil here possesses great fertility, and the whole valley teems with an exuberance of verdure.

July 18.—We changed our course this morning to the north, and passing up the valley of the creek, found a gap or pass in the first chain of mountains, through which, after much difficulty, we succeeded in forcing our wagons. This gap, although not very elevated, was broken up into deep and narrow gorges, filled with the angular debris of the adjoining heights, over which it required great care and patience to pass our train in safety. We, however, finally succeeded in reaching the open prairie upon the north, and found ourselves on the banks of a large stream, upon which we made our encampment. Our position is directly at the base of the most elevated mountain in the Witchita chain, which I have taken the liberty, in honor of our distinguished commanding general, to call "Mount Scott." This peak, towering as it does above all surrounding eminences, presents a very imposing feature in the landscape, and is a conspicuous landmark for many miles around. The altitude above the base, as determined by triangulation with the sextant, is eleven hundred and thirty-five feet.

To the north of Mount Scott lies one of the most beautiful and romantic valleys that I have ever seen. It is about three miles wide, enclosed between two ranges of the mountains, and through its centre winds a lovely stream of pure water, fifty yards wide and two feet deep, the lively current of which rushes wildly down over an almost continuous succession of rapids and rocky defiles. It is fringed upon each side with gigantic pecan, overcup, (*Quercus macrocarpa*,) white-ash, (*Fraxinus Americana*,) river-elm (*Ulmus nemoralis*,) and hackberry trees, (*Celtis*.) About the base of the mountains we find an abundance of post-oak, (*Quercus obtusiloba*,) and towards the summits, the red cedar (*Juniperus Virginiana*) grows.

The soil in this valley is highly productive, and sustains a heavy vegetation. The grass is very dense, of a good quality, and from two to three feet high; and were it not for the large flies that continue to phlebotomize our animals, they would luxuriate here.

Towards sundown I took my rifle, and, mounting a small Indian pony belonging to my negro servant, started up the creek for the purpose of hunting deer. After I had gone about two miles from camp, I suddenly discovered a buffalo bull very quietly cropping the grass under some oak trees near the creek. No sooner, however, did I see him, than, raising his head and giving one look in the direction I was approaching, he set off at a spanking gallop over the prairie. I applied the rowels most vigorously to the diminutive beast which I bestrode, and endeavored, by making a cut-off over the hills, to get within rifle range; but after exhausting all the efforts of the pony, I only found myself within about two hundred yards of the buffalo, and gave him a running salute as he passed, but did not observe him falter or make the slightest diminution in his speed; whereupon I reluctantly abandoned the chase and returned to camp.

CHAPTER VIII.

OLD INDIAN VILLAGES—BEAUTIFUL SCENERY—TRAP FORMATION—LOST MULE—BEAVER CREEK—PRAIRIE GUIDES—RUSH CREEK—WITCHITA AND WACO VILLAGES—MEXICAN PRISONERS—TALK WITH THE INDIANS—CROSS TIMBERS—KICKAPOOS—STRIKE WAGON TRACK—ARRIVAL AT FORT ARBUCKLE.

July 19.—At daylight this morning we crossed the creek after having excavated a passage for the wagons in the high banks, and travelled down the valley along the outer border of the timber in the bottom. The country over which we marched was of a similar character to that described about our last camp, and equally beautiful. We passed two old Indian villages, which John Bull, one of the hunters, says were formerly occupied by the Witchitas and Keechis; several of the lodges were still standing, with their old corn-fields near by.

Our camp is upon the creek about a mile above the village last occupied by the Witchitas before they left the mountains. Here they lived and planted corn for several years, and they have exhibited much taste and judgment in the selection of the site for their town. It is situated at the eastern extremity of the mountains, upon a plateau directly along the south bank of the creek, and elevated about a hundred feet above it, commanding an extended view of the country towards the north, south, and east. From its commanding position it is well secured against surprise, and is by nature altogether one of the most defensible places I have seen.

The landscape which is here presented to the eye has a most charming diversity of scenery, consisting of mountains, wood-lands, glades, water-courses, and prairies, all laid out and arranged in such peculiar order as to produce a witching effect upon the senses.

This must have been a favorite spot for the Indians; and why they have abandoned it I cannot imagine, unless it was through fear of the Comanches. It is only two years since they removed from here, and their lodge-frames are still standing, with the scaffolds upon which they dried their corn.

The soil, in point of fertility, surpasses anything we have before seen, and the vegetation in the old corn-fields is so dense, that it was with great difficulty I could force my horse through it. It consisted of rank weeds, growing to the height of twelve feet. Soil of this character must have produced an enormous yield of corn. The timber is sufficiently abundant for all purposes of the agriculturist, and of a superior quality. Most of the varieties of hard wood, such as overcup, post-oak, black-walnut, pecan, hackberry, ash, black or Spanish oak, (*Quercus elongata*,) elm, and china, besides cotton-wood and willow, are found here. We also found the wild passion flower, (*Passiflora incarnata*,) and a beautiful variety of the sensitive plant which we have not met with before.

Directly opposite the village, upon the north side, there is a large body of timber which extends across to the eastern branch of Cache creek; this unites with the branch upon which we are encamped, about a mile below the village.

Upon the south bank of the creek there is an immense natural meadow, clothed with luxuriant grasses, where hay might be procured sufficient to subsist immense numbers of cattle. Opposite our camp the creek flows directly at the base of a perpendicular wall of porphyritic trap, three hundred feet high, studded with dwarf cedars, which, taking shallow root in the crevices of the formation, receive their meagre sustenance from the scanty decomposition of the rocks. This escarpment has a columnar structure, with the flutings parallel and traversing the face in a vertical direction from top to bottom, and has the appearance of being the vertical section of a round hill that has been cleft asunder and one half removed, there being no appearance of a continuation of the formation upon the opposite bank of the creek. All the sides of this hill, except that upon the creek, are smooth, with gentle and easy slopes, covered with grass up to the very verge of the acclivity. On riding up the smooth ascent of this eminence, and suddenly coming upon the edge of the giddy precipice, one involuntarily recoils back with a shudder at the appearance of this strange freak of nature. Large veins of quartz were seen traversing this formation, and upon an examination of specimens we found it to be cellular or spongy, with the cells filled with liquid naphtha of about the consistence of tar, and having a strong resinous odor.

We have now reached the eastern extremity of the Witchita chain of mountains, and shall to-morrow morning cross the main creek below the village and strike our course for Fort Arbuckle, this being the nearest military post, and in our course for Fort Smith.

The more we have seen of the country about these mountains, the more pleased we have been with it. Indeed, I have never visited any country that in my opinion possessed greater natural local advantages for agriculture than this. Bounteous nature seems here to have strewed her favors with a lavish hand and to have held out every inducement for civilized man to occupy it. The numerous tributaries of Cache creek flowing from granite fountains, and winding like net-work in every direction through the valleys in the mountains—with the advantages of good timber, soil and grass, the pure, elastic, and delicious climate, with a bracing atmosphere—all unite in presenting rare inducements to the husbandman. It would only be necessary for our practical farmers to visit this locality: they could not be otherwise than pleased with it; and were it not for the fact that the greater part of the most desirable lands lie east of the 100th meridian of longitude, and within the limits of that vast territory ceded by our government to the Choctaws, it would be purchased and settled by our citizens in a very few years. As it is now situated far beyond the limits of the settlements, and directly within the range of the Comanches, it is of no use to the Choctaws themselves, as they do not venture among the prairie tribes, and do not even know the character of this part of their own territory. They have a superabundance of fertile lands bordering upon the Red and Canadian rivers, near the white settlements of Texas and Arkansas, and they prefer occupying those to going further out. They have thrown aside their primitive habits in a great degree, and abandoned the precarious and uncertain life of the hunter for the more quiet avocation of the husband- man. They look upon the wild Indian in much the same light as we do, and do not go among them; indeed, there is but little in common with them and the wild Indians.*

In consequence of losing one of our mules last night, we were de- tained later than usual this morning. Two of the Delawares went out at day-light in search of it, but returned in about two hours, not having been able to strike the track. We had up to this time been so fortunate as to lose no animals. I was therefore particularly desirous that the lost mule should be recovered, and intimated as much to our inter- preter, John Bushman, who had not joined in the first search. At the same time, I asked him what he thought were the chances of success. He replied, in his laconic and non-committal style, "I think maybe so find um—maybe not." I directed him to make an effort, and not give over the search as long as there remained the least prospect of

---

*The lands included within the Choctaw reservation, which are not occupied or made use of by them, are embraced within the 97th and 100th degrees of west longitude, and are bounded upon the north and south by the Canadian and Red rivers, being about one hundred and eighty miles in length by fifty in width, and constituting an aggregate of about nine thousand square miles of valuable and productive lands, or one thousand square miles more than the State of Massa- chusetts.

success. We then packed our wagons and started on towards Fort Arbuckle, crossing the creek below the old village, where it was forty yards wide and ten inches deep, with a rapid current flowing over a bed of gravel.

Upon the east bank of the creek we passed over a broad and level piece of bottom-land, covered with a dense crop of wild rice, and other rich grasses. We then left the valley in a course north of east, over the ridge dividing Cache from Beaver creek, until we reached a branch of the latter, upon which we encamped. The timber here is large and abundant; the water fresh, but standing in pools; and the soil good. I have crossed this same stream at four different places below here, and have invariably found the soil of a similar character and the timber large, consisting of pecan, elm, hackberry, oak, cotton-wood, and walnut, and generally confind to the borders of the stream.

Our most excellent and indefatigable hunter, John Bushman, returned this evening with the lost mule, having tracked him for twenty miles from where he left us. He had also killed a buffalo during the day, and brought us a piece of the hump. He states that from the time the mule left us until he overtook him he had continued to travel, without stopping, directly to the north, and at right-angles to the course we had been pursuing. I inquired of him if he did not become almost discouraged before he came up with the animal. He said no; that I had ordered him not to return without him, and that he should have been on the track yet if he had not overtaken him. I have no doubt such would have been the case, for he is a man of eminently determinate and resolute character, with great powers of endurance, and a most acute and vigilant observer, accompanied by prominent organs of locality and sound judgment. These traits of character, with the abundant experience he has had upon the plains, make him one of the very best guides I have ever met with. He never sees a place once without instantly recognising it on seeing it the second time, notwithstanding he may approach it from a different direction; and the very moment he takes a glance over a district of country he has never seen before, he will almost invariably point out the particular localities (if there are any such) where water can be found, when to others there seems to be nothing to indicate it. Such qualifications render the services of these people highly important, and almost indispensable in a tour upon the prairies.

An incident which was related to me as occurring with one of these guides a few years since, forcibly illustrates their character. The officer having charge of the party to which he was attached sent him out to examine a trail he had met with upon the prairie, for the purpose of ascertaining where it would lead to. The guide, after following it as far as he supposed he would be required to do, returned and reported that it led off into the prairies to no particular place, so far as he could discover. He was told that this was not satisfactory, and directed to take the trail again and to follow it until he gained the required information; he accordingly went out the second time, but did not return that day, nor the next, and the party, after a time, began to be alarmed for his safety, fearing he might have been killed by the Indians. Days and weeks passed by, but still nothing was heard of the guide, until, on arriving at the first border settlement, to their astonishment, he made his appearance among them, and, approaching the commanding officer said, "Captain, that trail which you ordered me to follow terminates here." He had, with indomitable and resolute energy, traversed alone several hundred miles of wild and desolate prairie, with nothing but his gun to depend upon for a subsistence, determined this time to carry out the instructions of his employer to the letter.

July 21.—We crossed two small branches this morning at four o'clock, and continued our course over undulating prairies, with smooth and even surfaces, frequently crossing small affluents of Beaver creek, where we found good running spring water, which can always be relied upon.

We had a copious shower this morning, which is the first rain that has fallen in several weeks.

There is good timber and grass upon all the branches we have passed to-day, and the soil is highly productive. We have also passed several groves of post-oak timber upon the ridges; this, however, for the most part, is small, short, and scrubby.

July 22.—Making an early start at two o'clock this morning, we ascended the eastern branch of Beaver creek to its source, when we found ourselves upon the ridge dividing this stream from Rush creek. The ridge is covered with timber similar to that of the Cross-Timbers, consisting of post-oak and black-jack, (*Quercus ferruginea*.)

Our road leads for five miles through this timber, when it emerges into a beautiful meadow, where the head of one of the branches of Rush creek takes its rise in large springs, and runs off in a fine bold stream, with a variety of hard timber along its borders. After following down this about two miles, we suddenly came in sight of several squaws who were collecting the tall grass which grows along the banks of the creek. They no sooner espied us than they jumped upon their horses and were about making off; most of them, however, stopped at the command of our interpreter, while one or two galloped away in the direction of the village to give notice of our approach. They proved to be Wacos and Witchitas, and informed us that their villages were about four miles in advance, at the same time inviting us to pay them a visit. We reached the villages (which were situated upon the banks of Rush creek) and encamped about half a mile below them in the valley.

Immediately on our arrival we were accosted by a large crowd of men, who were anxious to learn where we had been, and whether we had seen any Comanches; and as we were (I think) the first party of whites who had visited them at this place, they appeared very glad to see us—probably in anticipation of presents.

There are two villages here occupied by the Witchitas and Wacos respectively; they are situated in the rich and fertile valley of the creek, where they have cultivated corn, pumpkins, beans, peas, and melons. These people have no ploughs, or other agricultural instruments, but a small hoe, with which they prepare the ground for the reception of the seed, and do all other necessary work in its cultivation; yet the prolific soil gives them bountiful returns; and were it not for their improvident natures, they might, with little labor, have sufficient for the whole year. Instead of this, they only care for the present, and from the time the corn is fit for roasting, are continually eating and feasting until it is gone. They are then obliged to depend upon the precarious results of the chase during the remainder of the year.

The village of the Witchitas has forty-two lodges, each containing two families of about ten persons. These lodges are made by erecting a frame-work of poles placed in a circle in the ground, with the tops united in an oval form, and bound together with numerous withes or wattles, the whole nicely thatched with grass; and when completed, it makes a very commodious and comfortable domicil. The interior arrangements are such, that every person has a bunk, raised from the ground and covered with buffalo-hides, forming a couch which is far from being uncomfortable. When seated around their fires in the centre of the lodges, they have an air of domestic happiness about them which I did not expect to find.

The lodges are about twenty-five feet in diameter at the base, twenty feet high, and in the distance have very much the appearance of a group of hay-stacks. With the exception of a few families that live upon the Canadian, the whole Witchita nation is concentrated at this place: their numbers do not exceed five hundred souls. They have during the early settlement of Texas given more trouble to the people upon the northern borders of that State than any other Indians. They have no regard for truth, will steal, and are wholly unworthy of the least confidence, and their vicious propensities are only kept in check now from fear.

Living, as they do, between the white settlements and the prairie tribes, they are at the mercy of both; they seem to be conscious of this fact, and express a desire to be on terms of friendship with all their neighbors. At my urgent request they presented us with several bushels of green corn this evening, which was very acceptable, as we had seen no vegetables for several months.

The Wacos live about a mile above the Witchitas, in a village constructed precisely like the other. There are twenty lodges in this village, and about two hundred souls; their habits and customs are similar to the Witchitas, with whom they frequently intermarry, and are upon the best and most friendly terms.

Both of these tribes subsist for a great portion of the year upon buffalo and deer, and wear the buffalo robes, like the Comanches. They also use the bow and arrow for killing game; some of them, however, are provided with rifles, and are good shots. They have a large stock of horses and mules, many of which are the small Spanish breed with the Mexican brand upon them, and have probably been obtained from the prairie tribes; while others are large, well-formed animals, and have undoubtedly been stolen from the border white settlers.

We learned from the Witchitas, much to our surprise, that a report had been made to the commanding officer at Fort Arbuckle, by a Keechi Indian, to the effect that our whole party had been overpowered and massacred by the Comanches near the head of Red river. This information must have originated with the Comanches or Kioways, as they are the only tribes inhabiting the country about the sources of the river; neither the Keechies nor the Witchitas ever venture as far out into the plains as we have been.

The account given by the Indian was so circumstantial and minute in every particular, showing a perfect knowledge of all our movements, with our numbers and equipment, that the information was evidently communicated by persons who were near us at the time, and observing our movements. This accounts for the fact of their avoiding us upon all occasions, although we saw them several times, as has been observed, and frequently passed their camps that had been abandoned but a short time, yet they never came to us or communicated with us. They probably regarded us as out upon a hostile expedition, going into their country to chastise them for their depredations, and may have supposed that the report of our having been massacred would deter other parties from following us.

The old chief of the Witchitas (To-se-quash) informed us that Pah-hah-en-ka's band of the "Middle Comanches," in consequence of some of their people having been killed near one of the military posts in Texas, were much exasperated, and had burnt up the testimonials of good character given to them by United States authorities. They had always before preserved these papers with great care, and manifested much pride and satisfaction in exhibiting them to strangers. To-se-quash says they are now "very mad," and will fight us whenever they meet us.

July 23.—As it rained during the night, and still continues, we did not move forward to-day. During the morning I sent for the chiefs of the two villages, for the purpose of endeavoring to persuade them to surrender to me two Mexican prisoners in their possession : one a man about forty years of age, and the other a boy of fifteen. The man stated that he had been with the Witchitas since he was a child, and he was not now disposed to leave them; that he had become as great a rascal as any of the Indians, (which I gave full credence to,) and should not feel at home anywhere else.

It appeared, however, that the boy had only been with them a few months. He states that he was kidnapped by the Kioways from his home near Chihuahua; that in consequence of their brutal treatment he escaped and made his way to the Witchita mountains, where a Witchita hunter found him in nearly a famished state, and brought him to this place. He says he has been kindly treated by the Witchitas, but is anxious to leave them and go with us. He appears to be very intelligent, and reads and writes in his own language.

In a talk with the chiefs, I told them that the American people were now on terms of friendship with the Mexicans, and in a treaty we had obligated ourselves to return to them all prisoners in the hands of Indians in our territory, and to prevent further depredations being committed upon them; that the principal chief of the whites (the President) would not regard any tribe of Indians as friends who acted in violation of this treaty; that he confidently hoped and expected all the tribes who were friendly to our people would comply strictly with the requirements of the treaty, and give up all prisoners in their possession. I then requested them to release to me the boy, and told them if they did this I should make them some presents of articles that had been sent out by the President for such of his red children as were his friends. They hesitated for a long time, stating that the boy belonged to a Waco, and he loved him so much, that it was doubtful if he could be persuaded to part with him. Whereupon I told them that if they released the boy quietly, I should reward them; but otherwise I had determined to take him from them by force, and if compelled to resort to this course, should give them nothing in return. This appeared to have the desired effect, and they said if I would make the family into which he had been adopted a few presents, in addition to what I had promised them, they would release him. I accordingly distributed the presents, and took possession of the boy. Upon turning him over to us they divested him of the few rags of covering that hung about his person and reluctantly gave him to us, and he makes his exit from the Witchita nation in the same costume in which he entered the world. We soon had him comfortably clothed, and he is much delighted with the change. Captain McClellan will take him to San Antonio, from which place he will communicate with his relatives.

July 24.—We left the Witchita village at 4 o'clock this morning, and intended to have followed the trail which the Indians travel to Fort Arbuckle, but soon discovered that it crossed numerous brooks running through deep gullies impassable for wagons, which made it necessary for us to turn south towards the dividing ridge between Rush creek and Wild Horse creek. We followed this ridge for seven miles, and encamped upon a small affluent of Wild Horse creek. In our march to-day we passed over an elevated, waving country, interspersed with groves of oak. Upon each side of the dividing ridge are numerous small spring branches, flowing off to the right and left, and upon these there is an abundance of good timber, with soil of the best quality. We have passed the range of the grama grass, but still find the mezquite and other varieties of wild grasses, upon which our animals continue to thrive, and keep in excellent condition. After we had proceeded some ten miles upon our march this morning, we discovered that our friends the Witchitas had, in the characteristic style of their hospitality, abstracted from one of our wagons several articles which they probably supposed would be more useful to them than to us. Unfortunately, we were too far from the village to admit of going back and making them restore the articles. Our Spanish boy states that before he left, they advised him to seize the first opportunity that should offer to steal one of our horses, and make his escape to them.

July 25.—Our wagons were packed, and we were in motion at about 3 o'clock this morning, in a course nearly due east, down the right bank of Wild Horse creek for eight miles, when we entered the Cross-Timbers upon the ridge dividing this stream from Mud creek, (an affluent of Red river which puts in above the Washita.) Our encampment this evening is upon the border of a ravine in the timber, where we find good water and grass.

In our march to-day we have passed the heads of several branches running into Wild Horse, Beaver, Rush, and Mud creeks, upon all of which there is an exuberant vegetation, denoting a fertile soil. The timber is abundant and of a good quality, and the water, issuing from springs, is perennial. I have passed through the Cross-Timbers at five different points before this, and have always found them similar in character and composition.

Some Kickapoo hunters came into camp this evening, and we could not but remark the striking contrast between them and the Witchitas. They were fine-looking, well-dressed young men, with open, frank, and intelligent countenances, and seemed to scorn the idea of begging; while the others, as has been observed, are incessantly begging every article they see, and do not possess the slightest gratitude for favors received.

July 26.—At daylight this morning we resumed our march through the Cross-Timbers, keeping the dividing ridge for two miles, when we turned to the left and passed down near Wild Horse creek; but we found small streams, with abrupt banks, crossing our course so frequently that we had much difficulty in making progress. We, however, by hard labor in digging down banks and cutting through dense thickets, succeeded in making eight miles, and encamped upon a small spring branch in the Cross-Timbers. A short distance before we reached our present position we fell into an old Indian trail, where some wagons had passed several years before. We noticed where several small trees had been cut, and where the bark had been scraped off from others by the ends of the axles as they passed along.

July 27.—As soon as it was sufficiently light to enable us to see the trail this morning, we started on, keeping the old wagon trace through the timber for eight miles, when it led us into a road I had made the last season, between Fort Arbuckle and Fort Belknap, at a point fourteen miles from the former post. As soon as the men came in sight of this, they gave a prolonged and simultaneous shout of joy; it seemed to them like greeting an old familiar acquaintance: it was the first place they had recognised in several months, and it brought them near home.

The axes and spades were laid by in the wagons, as our labors in road-making terminate here; and I have no doubt the command are heartily rejoiced upon the occasion, as their duty since we left the Witchita mountains has been very laborious. Two miles after striking the road we emerged from the Cross-Timbers, and passing over a range of low mountains lying south of Wild Horse Creek valley, encamped nine miles from Fort Arbuckle.

July 28.—At one o'clock this morning we were upon the road again, and at daylight marched into Fort Arbuckle, where we found our friends much astonished and delighted at our sudden appearance among them, when they had supposed us all massacred by the Comanches. We are much indebted to the kind hospitality of the officers stationed here for the generous supply of vegetables with which they furnished one entire command during our stay with them. After an exclusive diet of meat and bread for several months, we could not have had a more welcome present than the fine fresh vegetables which their gardens afforded.

I shall remain here for two or three days to dispose of the stores on our hands, recruit our animals, and get the company in readiness to return to its station at Fort Belknap, under charge of Lieutenant Updegraff.

I feel a sincere regret at parting with the company, as the uniform good conduct of the men during the entire march of about a thousand miles merits my most sincere and heartfelt approbation. I have seldom had occasion even to reprimand one of them. All have performed the arduous duties assigned them with the utmost alacrity and good will; and when (as was sometimes the case) we were obliged to make long marches, and drink the most disgusting water for several days together, instead of murmuring and making complaints, they were cheerful and in good spirits. I owe them, as well as the officers and gentlemen who were with me, my most hearty thanks for their cordial co-operation with me in all the duties assigned to the expedition. It is probably in a great measure owing to this harmonious action on the part of all persons attached to the expedition, that it has resulted so fortunately.

We have lost no men by death, and, with the exception of the two cases of scurvy, there has been no sickness of consequence. And instead of any of our animals dying or straying away, we have had the especial good fortune of adding three horses, which we found upon the plains, to the number we received at the commencement of the march.

The animals, and particularly the oxen, many of which were so poor when they left Preston as to be considered almost useless, have all returned in fine condition, and are now much better capable of performing service than when they came into our hands.

CHAPTER IX.

PROMINENT FEATURES OF RED RIVER—FLOODS—CHAIN OF LAKES—CROSS-TIMBERS—ARABLE LANDS—ESTABLISHMENT OF A MILITARY POST UPON RED RIVER RECOMMENDED—ROUTE OF COMANCHES AND KIΘWAYS IN PASSING TO MEXICO—WAGON-ROUTE FROM FORT BELKNAP TO SANTA FE—NAVIGATION OF RED RIVER—ERRONEOUS OPINIONS IN REGARD TO RED RIVER—EXTENSIVE GYPSUM RANGE—EL LLANO ESTACADO.

In a comprehensive review of the physical characteristics of the particular section of Red river which is comprised within the limits of the district assigned to the attention of the expedition, it will not perhaps be considered irrelevant to make a few general observations upon the more prominent features of the country bordering upon this stream, from its confluence with the Mississippi to its sources. It will be observed, by reference to a map of the country embracing the basin of this river, that in ascending from the mouth, its general direction as high as Fulton, Arkansas, is nearly north and south; that here it suddenly changes its course and maintains a direction almost due east and west to its sources. One of the first peculiarities which strikes the mind on a survey of the topography of this extensive district of country, is the general uniformity of its surface: with the exception of the Witchita range no extensive chains of lofty mountains diversify the perspective, and but few elevated hills rise up to relieve the monotony of the prospect. Another distinguishing feature of this river is, that the country on its upper waters differs in every respect from that in the vicinity of its mouth. The valley is found to comprise two great geographical sections, each having physical characteristics entirely distinct from the other. The main branch of the river from the point where it debouches out of the Staked Plain, flows through an arid prairie country almost entirely destitute of trees, over a broad bed of light and shifting sands, for a distance, measured upon its sinuosities, of some five hundred miles. This country for the most part is subject to periodical seasons of drought, which preclude the possibility of cultivation except by means of artificial irrigation. It then enters a country covered with forest-trees of gigantic dimensions, growing upon an alluvial soil of the most pre-eminent fertility, which sustains a very diversified sylva, and affords to the planter the most bountiful returns of all the products suited to this latitude. On entering this section of the river we find that the borders contract, and the water, for a great portion of the year, washes both banks, at a high stage carrying away the loose alluvium from one side and depositing it upon the other in such a manner as to produce constant changes in the channel and to render the navigation difficult. This character prevails through the remainder of its course to the Delta of the Mississippi, and throughout this section it is subject to heavy inundations, which often flood the bottoms to such a degree as to produce very serious consequences to the planters, destroying their crops, and, upon subsiding, occasionally leaving a deposite of white sand over the surface, rendering it thenceforth entirely barren and worthless.

Below the great raft a chain of lakes continues to skirt the river for more than a hundred miles: these are supposed to have been formed in the ancient channels and low grounds of former streams, whose discharge had gradually been obstructed by an embankment formed of the sedimentary matter brought down the river from above.

These lakes are from five to fifty miles in length, from a quarter to three miles wide, and are filled and emptied alternately as the floods in Red river rise and fall: they serve as reservoirs, which in the inundation of the banks of the river receive a great quantity of water, and, as it subsides, empty their contents gradually, thereby tending to impede the rapid discharge of the floods upon the Delta. Like all rivers of great length which drain a large extent of country, Red river is subjected to periodical seasons of high and low water. The floods occur at very uniform epochs, but the quantity and elevation of the water, as well as its continuance at a high stage, vary constantly.

During the winter the water often remains high for several months, but the heavy rise which has almost invariably been observed during the month of June, often subsides in a very few days.

The geographical position of the sources of Red river being in latitude 34° 42′ and longitude 103° 7′ 10,″ and its confluence with the Mississippi in latitude about 31° and longitude 91° 50,′ it extends over three and a half degrees of latitude and eleven degrees of longitude. The barometrical elevation of its sources above the sea is twenty-four hundred and fifty feet. The estimated distance by the meanderings of the stream from the mouth to Preston, Texas, is sixteen hundred miles, and from this point to the sources of the main branch five hundred more, making the entire length of the river two thousand one hundred miles.

On emerging from the timbered lands upon Red river into the great plains, we pass through a strip of forest called the Cross-Timbers.

This extensive belt of woodland, which forms one of the most prominent and anomalous features upon the face of the country, is from five to thirty miles wide, and extends from the Arkansas river in a southwesterly direction to the Brazos, some four hundred miles.

At six different points where I have passed through it, I have found it characterized by the same peculiarities; the trees, consisting principally of post-oak and black-jack, standing at such intervals that wagons can without difficulty pass between them in any direction. The soil is thin, sandy, and poorly watered. This forms a boundary-line, dividing the country suited to agriculture from the great prairies, which, for the most part, are arid and destitute of timber. It seems to have been designed as a natural barrier between civilized man and the savage, as, upon the east side, there are numerous spring-brooks, flowing over a highly prolific soil, with a superabundance of the best of timber, and an exuberant vegetation, teeming with the delightful perfume of flowers of the most brilliant hues; here and there interspersed with verdant glades and small prairies, affording inexhaustible grazing, and the most beautiful natural meadows that can be imagined; while on the other side commence those barren and desolate wastes, where but few small streams greet the eye of the traveller, and these are soon swallowed up by the thirsty sands over which they flow: here but little woodland is found, except on the immediate borders of the water-courses.

From the point where Red river leaves the timbered lands, the entire face of the country, as if by the wand of a magician, suddenly changes its character. The bluffs now approach nearer the river, and the alluvial bottoms, which below here have been exceedingly rich and productive, contract, and do not support that dense and rank vegetation which characterizes the lower portion of the valley. The undergrowth of cane-brakes and vines disappears, and is no more seen throughout the entire extent of the valley. The lands adjacent gradually rise, and exhibit broad and elevated swells of surface, with spacious valleys intervening, and the soil continues to become more and more sterile as we ascend, until we reach the 101st degree of longitude, when from this point, with few exceptions, there is no more arable land.

Previous to my departure upon the expedition, I had been led to believe, from the representations of the Indians and others, that after passing Cache creek, no more good timber or land suited to cultivation would be met with upon the waters of Red river; but in this (as will have been observed) I was greatly in error, as we found much good timber and fertile land above this point.

The country drained by the numerous branches of Cache creek alone is very large, and possesses, in a remarkable degree, all the elements necessary for constituting a rich and productive agricultural district.

Including the valleys embraced within the Witchita mountains, there are upon a very moderate estimate, at least from seventy-five to eighty thousand acres of tillable lands upon the waters of this stream. In the valley of Otter creek there are also several thousand acres of rich alluvial lands, with timber in abundance; and upon Elk, Sweetwater, and the other small affluents of the North Fork, much land is found which would rank with our government surveyors as "first rate" in quality. All these would make up an aggregate of at least one hundred and fifty thousand acres of land, upon which cotton, corn, and most other grains, could be produced abundantly.

Could they be persuaded to lay aside their wandering habits and cultivate the soil, the amount of land here alluded to would be more than sufficient to sustain all the natives inhabiting this section of country; and the luxuriant and nutritious grasses which everywhere abound throughout the entire extent of the river basin, would furnish an inexhaustible amount of forage and grazing for their numerous animals. The winters here are mild, and it is seldom that the snow covers the ground more than a day or two at a time. There is a constant supply of good running-water upon all the minor tributaries to the North Fork, and sufficient woodland to supply farmers with fuel for a great number of years.

The soil in the valley of the main trunk of the river, as well as upon the Salt Fork, is thin and sandy, with very little timber or palatable water; and the country here possesses but few of the requisites essential to agriculture.

The Comanches and Kioways resort in great numbers to the waters of the north fork of Red river, where they find forage for their animals abundant during the winter months. Vestiges of their camps were everywhere observed along the whole course of the valley, from the Witchita mountains to the sources; and the numerous remains of the stumps of trees which had been cut down by them at different periods, indicated that this had been a favorite resort for them during many years. In several places we found camps that had only been deserted but a few days, and some where the fires were still burning. From the great extent of surface upon which the grass was cropped at some of these camping-places, and from the multitude of tracks still remaining, we inferred that they were supplied with immense numbers of animals; and they are undoubtedly attracted here by the superior quality of the grass, and the great abundance of cotton-wood which is found along the borders of the streams, upon the bark of which they fatten their favorite horses in the winter season.

Should the government authorities ever have occasion to communicate with these Indians, I have no doubt that many of them can always be found during the autumn, winter, and spring months along this branch of Red river; during the summer they leave and travel north in pursuit of the buffalo, generally ranging between the north fork of the Canadian and the Arkansas river.

We observed but few places upon the main branch of the river where the Indians had made their encampments. We, however, saw trails where they are accustomed to travel, crossing this branch and leading south towards the Brazos; indeed, a party with about fifty horses and mules had travelled along the bed of the Ke-che-a-qui-ho-no, through the gorge to the head of the river, but a short time previous to our passing.

The military posts already established upon the southwestern borders of Texas, with the two occupied by the fifth infantry in the direction of the headwaters of the Brazos, undoubtedly exercise a good influence over the southern Comanches who frequent that section; but there is a vast tract of country to the north of this, extending across Red river and the Canadian to the Arkansas, where there is no military post until reaching Fort Atkinson, upon the Santa Fé trace. Fort Arbuckle and Fort Scott are near the settlements, and they are now entirely out of the range of the prairie tribes. The northern and middle Comanches and the Kioways occupy this country, and go and come when and where they choose without the knowledge of any of our military authorities. These Indians probably commit more depredations upon the northern provinces of Mexico than any others. In passing back and forth upon these forays, they were formerly in the habit of taking a route crossing the Brazos and Colorado rivers, in the vicinity of some of the military posts in Western Texas; but since they have become acquainted with the localities of these posts, I have been informed by the Indians that they were so much harassed by the troops as to cause them to change their route; and now they generally pass to the north and west, entirely around this chain of posts.

It is a well-known fact, that whenever depredations have been committed by the Indians along the western borders of Texas, the perpetrators have almost invariably come from the north and returned in that direction; and when pursued, their trace has generally been found to lead towards Red river, in the direction of the western extremity of the Witchita chain of mountains. Such was the fact in the recent instances where animals were stolen from the posts upon the Brazos, and I cannot but believe, if there had been a garrison at some point upon Red river in the vicinity of the mountains, that the stolen animals might in a majority of cases have been recovered, and the authors of the depredations detected. Heretofore the troops stationed upon the Brazos, when sent in pursuit of Indians who had stolen animals, have followed them until their provisions were consumed, and have then been obliged to abandon the trail and turn back before coming near them; whereas, if they had started out from a post upon Red river, they would probably have been enabled to carry provisions sufficient to have served them until they could have reached the encampments where the freebooters had left their families.

A garrison established near the western extremity of the Witchita range of mountains would be in the heart of the Comanche country, and near the point where they cross Red river upon their marauding expeditions into Texas and Mexico.

The military authorities stationed here would have an opportunity of becoming acquainted with the chiefs, and with the character and habits of the Indians frequenting this section, and would have greater facilities for gaining their confidence and removing the unfavorable impression which they have heretofore entertained towards Americans. Believing that our government contemplates taking their hunting-grounds from them, they have always been suspicious of the motives of the whites who have visited their country; so much so, that upon one occasion they massacred a party of twenty men who attempted to survey a tract of land in Western Texas. They desire, therefore, to remain as far as possible away from the white settlements.

If troops were quartered in their country anywhere in the vicinity of the point I have mentioned, the Indians would by degrees become familiarized to their presence, and in time learn that instead of doing them injustice, the policy of our government towards them is such as would ultimately conduce to their welfare and prosperity.

At almost any point throughout the Witchita mountains, all the requisites for building and sustaining a military post are found in great profusion. The quality of the timber, soil, and water, are all far superior to that near the posts upon the Brazos river; and I firmly believe there is no more salubrious climate in the universe.

In my humble judgment, in view of what has been said, a military post established in the vicinity of these mountains, and garrisoned by a force of sufficient strength to command the respect of the Indians, would add more to the efficiency of the army in checking their depredations than any other position that is now occupied by the troops in Western Texas. This post would be about one hundred and forty miles distant from Fort Arbuckle; two hundred miles from Fort Washita; and one hundred and twenty from Fort Belknap; and being near Red river, (which it is believed will prove navigable, at a good stage of water, nearly as high as this point,) the troops could probably be furnished with supplies at a lower rate than at any of the military posts in this part of the country equi-distant from the seacoast.

Should it become necessary to march troops or transport supplies between the military posts upon the headwaters of the Brazos and Santa Fé, a better route cannot be desired for wagons than the one we have followed from Fort Belknap to the confluence of Cache creek, continuing up the north fork of Red river to near its source in the Staked Plain, and thence across in a northwesterly direction to the Canadian river, upon the south bank of which will be found a distinctly-marked wagon-trace, travelled by California emigrants in the summer of 1849, which leads in a very direct course, over firm and smooth ground, to Santa Fé. There is a bountiful supply of all that is essential to the comfort of the traveller and his animals upon this route; and good wood, water, and grass, are found so abundantly along the entire distance, that he need not make a single encampment without them all. The distance, measured along the route over which we travelled, from Fort Belknap to the mouth of Cache creek, is one hundred and twenty miles; from this point to the head of the north fork of Red river is two hundred and thirty-seven miles; to the Canadian, twenty-five miles; thence to Santa Fé, two hundred and ninety-five miles; making the aggregate distance between the termini six hundred and seventy-seven miles. These distances, as far as the Canadian, are measured upon the route over which we travelled in our explorations; and although its general course is reasonably direct, it is in some places circuitous, and could probably be shortened so as to reduce the distance to about six hundred and forty miles. The navigation of Red river with steamers of light draught is practicable at all times to Shreveport; and about four months of the year they have ascended without difficulty to Fort Towson.

During the past season, at a time when the river was at a low stage, a steamer drawing three and a half feet of water had no difficulty in ascending as high as Preston, near the confluence of the Washita. Several boats had previously reached this point upon the river; but as there are but few settlements above here, there has as yet been no inducements held out for boats to attempt the navigation of the river any higher. I am confident, however, from what I have seen of Upper Red river, that at a medium stage there will be sufficient depth of water for small steamers, such as ply upon some of the tributaries to the Mississippi, to ascend the river as high as where the two principal branches unite, (about fifty miles above the mouth of Cache creek.) As an evidence of this, on our outward march, at a time when the river was at a high stage, I had occasion for crossing frequently, but could find no place below the point mentioned where the water in the channel was of less depth than five feet; indeed, I do not think as many obstructions will be found above Preston as below, for the reason that there is but little woodland bordering upon the upper portion of the river, and consequently but few of those formidable obstacles called snags.

At a low stage the water in the river becomes very shallow, and can then be forded at any point. But during high water, the quicksands in the bed of the stream become loose and unstable, and make it hazardous to attempt a passage with animals. It was observed throughout that portion of the valley of the river which came under our observation, that it was bordered upon each side by three distinct terraces or benches, running parallel with the course of the stream. The first of these is from three to six feet high, from fifty to two hundred feet wide, and in places subject to overflow. The second, which is from ten to twenty feet above the first, is from two to five hundred feet wide, and is never submerged. The third varies from fifty to three hundred feet in elevation above the second, and forms the elevated line of bluffs that terminate the prairie lands adjacent to the valley.

In many places between the upper extremity of the Witchita mountains and the sources of the river, we found continuous chains of sand-hills, from twenty to fifty feet high, bordering the valley, and denuded of all herbage save a few plum-bushes and grape-vines. Although there is some good soil upon the small affluents to the main river, the country generally, immediately bordering it, is barren and sandy.

Several erroneous opinions have for many years been entertained in regard to the country upon the headwaters of Red river. For instance, it has generally been supposed, from the circumstance of a heavy rise occurring in the river during the month of June, at a time when there is generally no rain in the settlements, and during the dry season upon the plains, that the sources of the river would be found in lofty mountain ranges, where the melting of snows would account for the great amount of water passing through the channel at the season mentioned. But such is not the fact, as all the principal branches above Cache creek have their origin in the eastern borders of the table-lands of New Mexico, where there are no mountains. We, however, observed frequent and copious rains in the vicinity of the Witchita mountains during the season of the June flood; and I am of the opinion that here is the source whence much of the water is derived.

As the water in the river has a very bitter and disagreeable taste, it has been conjectured that it passed in its course through extensive salt plains: but this I also found to be an error. We saw no deposite of chloride of sodium in the vicinity of the river; the peculiar taste being communicated by ingredients that it receives in flowing for a hundred miles over a gypsum formation. An analysis of this water, under the direction of Dr. Clark, of Amherst College, gives the following results, from which it will be seen that the per-centage of salt is small:*

Weight of water in fluid ounces	4
" " water in fluid grammes	127.800
" " chlorine present	.051
" " lime	.033
" " sulphuric acid	.095
Sulphates of soda and magnesia	.168

Regarding the lime as a sulphate, and the residue of sulphuric acid as united with magnesia, and the chlorine as united with the sodium, we have the following results:

Weight of sulphate of lime	.080
" " sulphate of magnesia	.073
" " chloride of sodium	.084
Weight of the whole	.237
Per-centage of matter in solution	19

This gypsum range forms an immense belt, which extends across the country for some four or five hundred miles. Col. Long speaks of seeing it upon the Arkansas; and I have myself passed through it at four other different points south of this, embracing a range of some three hundred miles. It is regarded by Dr. Hitchcock as the most extensive deposite of this mineral in North America. I have everywhere found it char-

---

*I have understood since our return that the Indians have recently discovered a deposite of salt (chloride of sodium) about three miles to the south of our return route, near the western extremity of the Witchita mountains.

acterized by the same peculiarities, with the water issuing from it invariably bitter and unpalatable.

The Arkansas, Canadian, Brazos, Colorado, and Pecos rivers, pass through the formation, and a similar taste is imparted to the waters of all. Several of these also have their sources in the same elevated table-lands as Red river, and where they make their exit from this plateau their beds are confined to vast sluices or cañons, the sides of which rise very abruptly to an enormous height above the surface of the water. The barren mesa, in which these streams take their rise, extends from the Canadian river, in a southerly course, to near the confluence of the Pecos with the Rio Grande, some four hundred miles, between the 32d and 37th parallels of north latitude. It is in places nearly two hundred miles in width, and is embraced within the 101st and 104th meridians of west longitude. The approximate elevation of this plain above the sea, as determined with the barometer, is two thousand four hundred and fifty feet. It is much elevated above the surrounding country, very smooth and level, and spreads out in every direction as far as the eye can penetrate, without a tree, shrub, or any other herbage to intercept the vision. The traveller in passing over it sees nothing but one vast, dreary, and monotonous waste of barren solitude. It is an ocean of desert prairie, where the voice of man is seldom heard, and where no living being permanently resides. The almost total absence of water causes all animals to shun it: even the Indians do not venture to cross it except at two or three points, where they find a few small ponds of water. I was told in New Mexico that, many years since, the Mexicans marked out a route with stakes across this plain, where they found water; and hence the name by which it is known throughout Mexico, of "El Llano Estacado," or the "Staked Plain."

CHAPTER X.

INDIANS OF THE COUNTRY—HABITS OF THE COMANCHES AND KIOWAYS—SIMILARITY BETWEEN THEM AND THE ARABS AND TARTARS—PREDATORY EXCURSIONS INTO MEXICO—WAR IMPLEMENTS—INCREDULITY REGARDING THE CUSTOMS OF THE WHITES—METHOD OF SALUTING STRANGERS—DEGRADED CONDITION OF THE WOMEN—AVERSION TO ARDENT SPIRITS—PRAIRIE INDIANS CONTRASTED WITH INDIANS OF THE EASTERN STATES—BUFFALOES—PROBABLE CONDITION OF THE INDIANS UPON THE EXTERMINATION OF THE BUFFALOES—PERNICIOUS INFLUENCES OF TRADERS—SUPERSTITIONS OF THE NATIVES.

The country over which we passed is frequented by several tribes of Indians, who follow the buffalo, and subsist almost exclusively upon the uncertain products of the chase. The Witchitas, Wacos, Kechies, and Quapaws, all resort to the country about the Witchita mountains, where a few years since they had their thatched villages and corn-fields, but they have recently removed near the white settlements. The Witchitas and Wacos, as before stated, are now living upon Rush creek, while the Kechies and Quapaws are upon Chouteau's creek, an affluent of the Canadian. The Witchitas and Kechies each number about one hundred warriors; the Wacos about eighty; and the Quapaws only about twenty-five. They all use the horse in their hunting and war expeditions, and are possessed of a good supply of these animals. The history of the Quapaws, a minute remnant of what was once a large and powerful nation of Indians, called the "Arkansas," but now only numbering a very few lodges of miserable half-starved beggars, is truly melancholy. Father Charlevoix, in his "Historical Journal of a Voyage down the Mississippi," speaks of visiting them, and found them at that time very numerous and warlike. He says of them: "The Arkansas, or Quapaws, are reckoned to be the tallest and best-shaped of all the savages of this continent, and they are called, by way of distinction, 'the fine men.'" He describes them as occupying at the time of his visit four villages, one of which was upon the Mississippi, a short distance above the mouth of the Arkansas. They were, according to him, composed of the confederated remnants of several ruined nations.

In the time of Du Pratz these Indians had all moved up the Arkansas, and were living about twelve miles from the mouth of White river; they were then quite numerous, and he compliments them by saying that they were no less distinguished as warriors than hunters, and that they were the first nation that succeeded in conquering the warlike and numerous Chickasaws. It is related that upon one occasion they encountered the Chickasaws, who, in consequence of having no powder, considered it most prudent to make a precipitate retreat; whereupon the Quapaw chief, understanding the cause, determined they should be placed on an equality, and ordered all his warriors to empty their powder-horns into a blanket, and making an equal division of the powder, he gave one-half to his enemies. The battle then commenced, and in a short time terminated with a signal defeat of the Chickasaws, who retreated with a loss of ten killed and five prisoners, while the Quapaws only lost one man. They were also distinguished for their friendship to the early settlers along the Arkansas river, and it is much to be deplored that this once numerous and valorous nation is so fast approaching annihilation. The two most numerous and powerful tribes of Indians frequenting the country upon Upper Red river are the Comanches and Kioways; the former range from the Witchita mountains to the sources of the river, while the latter occasionally visit the headwaters, but seldom come as far down as the mountains. These tribes have similar habits, but speak different languages. The most numerous and warlike nation is that of the Comanches, who are separated into three distinct local grand divisions, namely: the Northern, Middle, and Southern; each of these is subdivided into several bands, commanded by separate chiefs.

The Northern and Middle Comanches subsist almost entirely upon the flesh of the buffalo; they are known among the other Indians as "buffalo-eaters," and are generally found at their heels, migrating with them from place to place upon those vast and inhospitable plains of the West, the greater portion of which are incapable of cultivation, and seem destined in the future, as in the past, to be the abode of the wandering savage, possessing, as they do, so few attractions to civilized man. This vast district, however, exhibits one characteristic which compensates for many of its asperities: perhaps no part of the habitable globe is more favorable to human existence, so far as the atmosphere is concerned, than this. Free from marshes, stagnant water, great bodies of timber, and all other sources of poisonous malaria, and open to every wind that blows, this immense grassy expanse is purged from impurities of every kind, and the air imparts a force and vigor to the body and mind which repays the occupant in a great measure for his deprivations. Nature, which almost everywhere exhibits some compensation to man for great hardships, has here conferred upon him health, the first and best of her gifts. It is a fact worthy of remark, that man, in whatever situation he may be placed, is influenced in his modes of existence, his physical and moral condition, by the natural resources of climate, soil, and other circumstances around him, over the operations of which he has no control. Fortunately, such is the flexibility of his nature that he soon learns to adapt himself to the hardest and most untoward circumstances, and, indeed, ultimately becomes not only reconciled to his lot, but persuades himself that his condition is far preferable to that of most others.

The example of our western-border settlers is illustrative of this fact, as they continue to remove farther and farther west as the settlements encroach upon them, preferring a life of dangerous adventure and solitude to personal security and the comforts and enjoyments of society; and what was at first necessity to them, becomes in time a source of excitement and pleasure.

The nomadic Indian of the prairies demonstrates the position still more forcibly: free as the boundless plains over which he roams, he neither knows nor wants any luxuries beyond what he finds in the buffalo or the deer around him. These serve him with food, clothing, and a covering for his lodge, and he sighs not for the titles and distinction which occupy the thoughts and engage the energies of civilized man. His only ambition consists in being able to cope successfully with his enemy in war, and in managing his steed with unfailing adroitness. He is in the saddle from boyhood to old age, and his favorite horse is his constant companion. It is when mounted that the Comanche exhibits himself to the best advantage: here he is at home, and his skill in various manœuvres which he makes available in battle—such as throwing himself entirely upon one side of his horse, and discharging his arrows with great rapidity towards the opposite side from beneath the animal's neck while he is at full speed—is truly astonishing. Many of the women are equally expert, as equestrians, with the men. They ride upon the same saddles and in the same manner, with a leg upon each side of the horse. As an example of their skill in horsemanship, two young women of one of the bands of the Northern Comanches, while we were encamped near them, upon seeing some antelopes at a distance from their camp, mounted horses, and with lassos in their hands set off at full speed in pursuit of this fleetest inhabitant of the plains. After pursuing them for some distance, and taking all the advantages which their circuitous course permitted, they finally came near them, and, throwing the lasso with unerring precision, secured each an animal and brought it back in triumph to the camp. Every warrior has his war-horse, which is the fleetest that can be obtained, and he prizes him more highly than anything else in his possession, and it is seldom that he can be induced to part with him at any price. He never mounts him except when going into battle, the buffalo chase, or upon state occasions. On his return from an excursion he is met at the door of his lodge by one of his wives, who takes his horse and attends to its wants with the utmost care. The prairie warrior performs no menial labor; his only occupation is in war and the chase. His wives, who are but little dearer to him than his horse, perform all the drudgery. He follows the chase, he smokes his pipe, he eats and sleeps; and thus he passes his time, and in his own estimation he is the most lordly and independent sovereign in the universe. Such are some of the characteristics of the prairie Indians; and I cannot dismiss the subject without remarking that, in addition to the physical similitude between the deserts of Arabia, the steppes of Central Asia, and the prairie *mesas* of our own country, a very striking resemblance is also observed in the habits and customs of the respective inhabitants. The Arabs of the desert, the Tartar tribes, and the aboriginal occupants of the prairies, are alike wanderers, having no permanent abiding-places, transporting their lodges wherever they go; and where these are pitched, there are their homes. They permit no authorities to control them but such as receive the unanimous sanction of the masses, and the rule of their leaders is guided by the counsels of their old men, who in many cases allay dissensions and curb the impetuosity of ambitious young warriors, whose thirst for fame would often involve the nation in protracted wars. Thus their government is patriarchal, guided by matured and fraternal counsels. They are insensible to the wants and comforts of civilization; they know neither poverty nor riches, vice nor virtue, and are alike exempt from the deplorable vicissitudes of fortune. Theirs is a happy state of equality, which knows not the perplexities of ambition nor the crimes of avarice. They never cultivate the soil, but subsist altogether upon game and what they can steal. They are alike the most expert horsemen in the world, and possess the same fond attachment for the animal. I once made an effort to purchase a favorite horse from a chief of one of the bands of the Southern Comanches, (Se-na-co,) and offered him a large price, but he could not be persuaded to part with him. He said the animal was one of the fleetest in their possession; and if he were to sell him, it would prove a calamity to his whole band, as it often required all the speed of this animal to insure success in the buffalo chase; that his loss would be felt by all his people, and he would be regarded as very foolish: moreover, he said, (patting his favorite on the neck,) "I love him very much."

The only property of these people, with the exception of a few articles belonging to their domestic economy, consists entirely in horses and mules, of which they possess great numbers. These are mostly pillaged from the Mexicans, as is evident from the brand which is found upon them. The most successful horse-thieves among them own from fifty to two hundred animals.

In their political and domestic relations there is also a similarity to the Old World nomads. They are governed by a chief, the tenure of whose office is hereditary so long as his administration meets the approbation of his followers. He leads them to war, and presides at their deliberations in council; but should he disgrace himself by any act of cowardice or mal-administration, they do not hesitate to depose him and place a more competent man in his stead. Their laws are such as are adapted to their peculiar situation, and are sanctioned by the voice of the people. Their execution is vested in the subordinate chiefs or captains, as they are called, and they are promptly and rigidly enforced. In respect to the rights of property, their code is strictly Spartan. They are perhaps as arrant freebooters as can be found upon the face of the earth; and they regard stealing from strangers as perfectly legitimate and honorable, and that man who has been most successful in this is the most highly honored by his tribe; indeed, a young man who has not made one or more of these expeditions into Mexico is held in but little repute. In evidence of this, I was told by an old chief of the Northern Comanches, called Is-sa-keep, that he was the father of four sons, who he said were as fine young men as could be found; that they were a great source of comfort to him in his old age, and could steal more horses than any other young men in his band.

As these forays are often attended with much toil and danger, they are called "war expeditions." It not unfrequently happens that but six or eight young men set out upon one of these adventures, and the only outfit they require is a horse, with their war equipments, consisting of the bow and arrows, lance and shield, with occasionally a gun. Thus prepared, they set out upon a journey of a thousand miles, or more, through a perfectly wild and desolate country, dependent for subsistence wholly upon such game as they may chance to find. They make their way to the northern provinces of Mexico, where they lie in wait near some hacienda until a favorable opportunity offers to sweep down upon a solitary herdsman, and, with the most terrific yells, drive before them all the animals they desire. Wo to the panic-stricken ranchero who fails to make a precipitate retreat, as they invariably kill such men as offer the slightest impediment to their operations, and take women and children prisoners, whom they hold in bondage of the most servile character. They are sometimes absent from their tribes two years or more before their success is sufficient to justify their returning with credit to themselves.

The use of the bow, which is the favorite arm and constant appendage of the prairie Indian, and which he makes use of exclusively in hunting the buffalo, is taught the boys at a very early age; and by constant and careful practice, they acquire a degree of proficiency in the art that renders them, when grown up to manhood, formidable in war, as well as successful in the chase. Their bows are made of the tough and elastic wood of the "bois d'arc," or Osage orange (*Maclura aurantiaca*,) strengthened and reinforced with the sinews of the deer wrapped firmly around them, and strung with a cord made of the same material. They are not more than one-half the length of the old English long-bow, which was said to have been sixteen hands' breadth in length. The arrows are twenty inches long, of flexible wood, with a triangular point of iron at one end, and two feathers, intersecting each other at right-angles, at the opposite extremity. At short distances the bow, in the hands of the Indian, is effective, and frequently throws the arrow entirely through the huge carcass of the buffalo. In using this instrument, the Indian warrior protects himself from the missiles of his enemy with a shield of circular form, covered with two thicknesses of hard, undressed buffalo-hide, separated by a space of about an inch, which is stuffed with hair: this is fastened to the left arm by two bands, in such a manner as not to interfere with the free use of the hand, and offers such resistance that a rifle-ball will not penetrate it unless it strikes perpendicular to the surface. They also make use of a war-club, made by bending a withe around a hard stone of about two pounds weight, which has been previously prepared with a groove in which the withe fits, and is thereby prevented from slipping off. The handle is about fourteen inches long, and bound with buffalo-hide.

The Comanche men are about the medium stature, with bright, copper-colored complexions and intelligent countenances, in many instances with aquiline noses, thin lips, black eyes and hair, with but little beard. They never cut the hair, but wear it of very great length, and ornament it upon state occasions with silver and beads. Their dress consists of leggins and moccasins, with a cloth wrapped around the loins. The body is generally naked above the middle, except when covered with the buffalo-robe, which is a constant appendage to their wardrobe. The women are short, with crooked legs, and are obliged to crop their hair close to their heads. They wear, in addition to the leggins and moccasins, a skirt of dressed deer-skin. They also tattoo their faces and breasts, and are far from being as good looking as the men.

Notwithstanding that these people are hospitable and kind to strangers, and apparently amiable in their dispositions, yet, when a warrior conceives himself injured, his thirst for revenge knows no satiety. Grave and dignified in his deportment, and priding himself upon his coolness of temper and the control of his passions, yet, when once provoked, he, like the majority of his race, is implacable and unrelenting; an affront is laid up and cherished in his breast, and nothing can efface it from his mind until ample reparation has been made. He has no idea of forgiveness: the insult must be atoned for by blood. With many tribes, quarrels can often be settled by presents to the injured party; but with the Comanches, their law of equity is of such a character that no reconciliation can take place until the reproach is wiped out with the blood of their enemy. They make no use of money except for ornaments. Like other tribes, they are fond of decking themselves with paint, beads, and feathers; and the young warrior often spends more time at his toilet than the most conceited coxcomb that can be found in civilized life. Bright red and blue are their favorite colors; and vermilion is an important article in the stock of goods of one of their traders. This they always carry about their persons; and whenever they expect to meet strangers, they always (provided they have time) make their toilet with care, and paint their faces. Some few of their chiefs who have visited their Great Father at Washington, have returned strongly impressed with the numerical power and prosperity of the whites; but the great majority of them being entirely ignorant of everything that relates to us, and the most of them having never even seen a white man, believe the Comanches to be the most powerful nation in existence; and the relation of facts which conflict with this notion, by their own people, to the masses of the tribes at their prairie firesides, only subjects the narrator to ridicule, and he is set down as one whose brain has been turned by the necromancy of the pale-faces, and is thenceforth regarded as wholly unworthy of confidence.

Having upon one occasion a Delaware and a Comanche with me in the capacity of guides, I was much diverted with a conversation which passed between them in my presence, and which was interpreted to me by the Delaware. It appeared that the latter had stated to the other the fact of the sphericity of the earth's surface. This idea being altogether new and incomprehensible to the Comanche, was received with much incredulity, and, after gazing a moment intently at the Delaware to ascertain if he was sincere, he asked if that person took him for a child, or if he looked like an idiot. The Delaware said no; but that the white people, who knew all about these things, had ascertained such to be the fact; and added, that the world was not only round, but that it revolved in its orbit around the sun. The Comanche very indignantly replied, that any man of sense could, by looking off upon the prairie, see at a glance that the earth was perfectly level; and, moreover, that his grandfather had been west to the end of it, where the sun disappeared behind a vertical wall. The Delaware continued, in his simple but impressive manner, to describe to the Comanche the steam-engine, with other objects of interest he had seen among the whites, all of which the Comanche regarded as the product of a fertile imagination, expressly designed to deceive him; and the only reply that he deigned to make was an occasional exclamation in his own language, the interpretation of which the Delaware pronounced to be, "Hush, you fool!" I then endeavored to explain to the Delaware the operation of the magnetic telegraph, and, in illustration of its practical utility, stated to him that a message could be sent a distance of one thousand miles, and an answer returned, in the short space of ten minutes time. He seemed much interested in this, and listened attentively to my remarks, but made no comments until I requested him to explain it to the Comanche, when he said, "I don't think I tell him that, Captain; for the truth is, I don't believe it myself."

The mode of life of the prairie tribes, owing to their unsettled and wandering habits, is such as to render their condition one of constant danger and apprehension. The security of their numerous animals from the encroachments of their enemies, and their constant liability to attacks, make it imperatively necessary for them to be at all times upon the alert. Their details for herdsmen are made with as much regularity as the guard-details at a military post; and even in times of the most profound peace, they guard their animals both night and day, while scouts are often patrolling upon the adjoining heights to give notice of the approach of strangers, when their animals are hurried to a place of security, and everything made ready for defence. The manner in which they salute a stranger is somewhat peculiar, as my own reception at one of their encampments will show. The chief at this encampment was a very corpulent old man, with exceedingly scanty attire, who, immediately on our approach, declared himself a great friend of the Americans, and persisted in giving me evidence of his sincerity by an embrace, which, to please him, I forced myself to submit to, although it was far from agreeable to my own feelings. Seizing me in his brawny arms while we were yet in the saddle, and laying his greasy head upon my shoulder, he inflicted upon me a most bruin-like squeeze, which I endured with a degree of patient fortitude worthy of the occasion; and I was consoling myself upon the completion of the salutation, when the savage again seized me in his arms, and I was doomed to another similar torture, with his head on my other shoulder, while at the same time he rubbed his greasy face against mine in the most affectionate manner; all of which proceeding he gave me to understand was to be regarded as a most distinguished and signal mark of affection for the American people in general, whom, as he expressed it, he loved so much that it almost broke his heart; and in particular for myself, who, as their representative, can bear testimony to the strength of his attachment. On leaving his camp, the chief shook me heartily by the hand, telling me at the same time that he was not a Comanche, but an American; and as I did not feel disposed to be outdone in politeness by an Indian, I replied, in the same spirit, that there was not a drop of Anglo-Saxon blood in my veins, but that I was wholly and absolutely a Comanche, at which he seemed delighted, duly understanding and appreciating the compliment. These people are hospitable and kind to all with whom they are not at war; and on the arrival of a stranger at their camps, a lodge is prepared for him, and he is entertained as long as he chooses to remain among them. They are also kind and affectionate to each other, and as long as anything comestible remains in the camp, all are permitted to share alike; but with these exceptions, they are possessed of but few virtues. Polygamy is sanctioned and is very common among them, every man being allowed as many wives as he can support.

Within the past few years the Comanches have (for what reason I could not learn) taken an inveterate dislike to the negroes, and have massacred several small parties of those who attempted to escape from the Seminoles and cross the plains for the purpose of joining Wild Cat upon the Rio Grande. Upon inquiring of them the cause of their hostility to the blacks, they replied that it was because they were slaves to the whites; that they were sorry for them. I suspect, however, that they were actuated by other motives than they cared about acknowledging, and that instead of wishing to better their condition by sending them to another world, where they would be released from the fetters of bondage, they were apprehensive, if they permitted them to pass quietly, that in time Wild Cat's followers upon the Rio Grande would augment to such a degree that he would interfere with their marauding operations along the Mexican borders. During the past year they have also been hostile towards the Delawares and Shawnees, and have killed several individuals who have been into their country in small parties. The Creek Indians, who exercise a good influence over the prairie tribes, have counselled them to commit no further acts of hostility upon these Indians, and I presume they will take measures to enforce a strict adherence to their wishes in this respect. These people, who are so extremely jealous of their own freedom that they will often commit suicide rather than be taken prisoners, are the more prone to enslave others, and this dominant principle is carried to the greatest extreme so far as regards their women. A beast of burden and a slave to the will of her brutal master, yet, strange as it may appear, the Comanche woman seems contented with her lot, and submits to her fate without a murmur. The hardships imposed upon the females are most severe and cruel. The distance of rank and consideration which exists between the black slave and his master is not greater than between the Comanche warrior and his wife. Every degrading office that is imposed upon the black by the most tyrannical master, falls, among the Comanches, to the lot of the wretched female. They, in common with other Indians, are not a prolific race; indeed, it is seldom that a woman has more than three or four children. Many of these, owing to unavoidable exposure, die young; the boys, however, are nurtured with care and treated with great kindness by their mothers, while the girls are frequently beaten and abused unmercifully. I have never seen an idiot, or one that was naturally deformed, among them.

Of all the Indians I had before encountered, there were none who had not an extreme fondness for spirituous liquors. The prairie tribes that I have seen, say the taste of such liquor is not pleasant; that it makes fools of them, and that they do not desire it. If there are exceptions to this, I think they may be set down as factitious rather than natural; the appetite having been created by occasional indulgence in the use of a little at a time.

The diet of these people is very simple; from infancy to old age their only food, with the exception of a few wild plants which they find on the prairies, is fresh meat, of which, in times of plenty, they consume enormous quantities. In common with many other tribes, they can, when necessity demands it, abstain from eating for several days without inconvenience, and they are enabled to make up at one meal the deficiency. All of them are extravagantly fond of tobacco, which they use for smoking, mixed with the dried leaves of the sumach, inhaling the smoke into their lungs and giving it out through their nostrils. Their language is verbal and pantomimic. The former consists of a very limited number of words—some of which are common to all the prairie tribes. The latter, which is exceedingly graceful and expressive, is the court language of the plains, and is used and understood with great facility and accuracy by all the tribes from the Gila to the Columbia; the motions and signs to express ideas being common to all. In contemplating the character of the prairie Indian, and the striking similarity between him and the Arab and Tartar, we are not less astonished at the absolute dissimilarity between these and the aboriginal inhabitants of the Eastern States. The latter, from the time of the discovery of the country, lived in permanent villages, where they cultivated fields of corn, and possessed strong attachment for their ancestral abodes and sepulchres: they did not use horses, but always made their hunting and war expeditions on foot, and sought the cover of trees on going into battle; while the former have no permanent abiding-places, never cultivate the soil, are always mounted, and never fight a battle except in the open prairie, where they charge boldly up to an enemy, discharge their arrows with great rapidity, and are away before their panic-stricken antagonist can prepare to resist or retaliate. In their treatment of prisoners of war there was also a very marked difference. The eastern tribes, although they put their prisoners to tortures of the most appalling character, seldom, if ever, violate the chastity of the females; while, on the contrary, the prairie Indians do not put their prisoners to death by prolonged tortures, but invariably compel the females to submit to their lewd embraces. There is at this time a white woman among the Middle Comanches, by the name of Parker, who, with her brother, was captured while they were young children, from their father's house in the western part of Texas. This woman has adopted all the habits and peculiarities of the Comanches; has an Indian husband and children, and cannot be persuaded to leave them. The brother of the woman, who had been ransomed by a trader and brought home to his relatives, was sent back by his mother for the purpose of endeavoring to prevail upon his sister to leave the Indians and return to her family; but he stated to me that on his arrival she refused to listen to the proposition, saying that her husband, children, and all that she held most dear, were with the Indians, and there she should remain. As the prairie Indians depend almost entirely on the buffalo for a subsistence and for clothing, it becomes a question of much interest, what will be the fate of these people when these animals shall have become extinct? Formerly, buffaloes were found in countless herds over almost the entire northern continent of America, from the 28th to the 50th degree of north latitude, and from the shores of Lake Champlain to the Rocky mountains. As it is important to collect and preserve all facts connected with the history of this interesting and useful animal before the species becomes extinct, I trust I shall be pardoned for introducing a few quotations from authors, touching their early history, which to me appear highly interesting. In a work published at Amsterdam in 1637, called "New English Canaan," by Thomas Morton, one of the first settlers of New England, he says: "The Indians have also made description of great *heards* of well-growne beasts that live about the parts of this lake (Erocoise,) now Lake Champlain, such as the Christian world (until this discovery) hath not *bin* made acquainted with. These beasts are of the bigness of a *cowe*, their flesh being very good *foode*, their hides good leather; their fleeces very useful, being a kind of *woole*, as fine almost as the *woole* of the beaver; and the *salvages* do make garments thereof. It is *tenne yeares* since first the relation of these things came to the *eares* of the English."

It is stated by another author, (Purchas,) that as early as in 1613 the adventurers in Virginia discovered a "slow *kinde* of *cattell* as *bigge* as kine, which were good *meate*."

The limit of the buffalo range on the north has been given differently by different writers. In a work published in London in 1589, by Hukluyt, it is stated, that in the island of New Foundland were found "*mightie beastes*, like to camels in greatness and their *feete* were cloven." He then says: "I did see them *farre* off, not able to *discerne* them perfectly, but their steps showed that their *feete* were cloven, and *bigger* than the *feete* of camels. I suppose them to be a kind of *buffes*, which I read to *bee* in the *countreys* adjacent, and very many in the *firme* land."

It is supposed by some that these animals may have been the musk-ox. They were found by Captain Franklin as high as 60° north latitude. Although it is doubtful whether the buffalo ever ranged beyond the Rocky mountains, yet they have been found as far west as the western slope. They formerly ranged free and uninterrupted over the boundless plains of the West, only guided in their course by that faithful instinct which invariably led them to the freshest and sweetest pastures. Their only enemy then was the Indian, who supplied himself with food and clothing from the immense herds around his door; but would have looked upon it as sacrilege to destroy more than barely sufficient to supply the wants of his family. Thus this monarch of the plains was allowed free range from one end of the continent to the other. But this happy state of things was not destined to continue: an enemy appeared, who made great havoc among them, and in a short time caused a very sensible diminution in their numbers, and much contracted the limits of their wanderings. This enemy was the white man, who, in his steady march, causes the original proprietor of the soil to recede before him, and to diminish in numbers almost as rapidly as the buffalo. Thousands of these animals were annually slaughtered for their skins, and often for their tongues alone; animals whose flesh is sufficient to afford sustenance to a large number of men are sacrificed to furnish a "bon bouche" for the rich epicure. This wholesale slaughter on the part of the white man, with the number consumed by the Indians, who are constantly on their trail, migrating with them as regularly as the season comes round, with the ravenous wolves that are always at hand to destroy one of them if wounded, gives the poor beast but little rest or prospect of permanent existence. It is only eight years since the western borders of Texas abounded with buffaloes; but now they seldom go south of Red river, and their range upon east and west has also very much contracted within the same time; so that they are at present confined to a narrow belt of country between the outer settlements and the base of the Rocky mountains. With this rapid diminution in their numbers, they must in the course of a very few years become exterminated. What will then become of the prairie Indian, who, as I have already remarked, relies for subsistence, shelter, and clothing, on the flesh and hide of this animal? He must either perish with them, increase his marauding depredations on the Mexicans, or learn to cultivate the soil. As the first law of our nature is self-preservation, it is not probable that he will sit down and quietly submit to starvation; he must, therefore, resort to one of the latter alternatives. But as he has no knowledge of agriculture, considers it the business of a slave, and very much beneath the dignity of a warrior, it appears reasonable to suppose that he will turn his attention to the Mexicans, over whom he has held the mastery for many years. Heretofore he has plundered these people to supply himself with animals for his own use and for traffic.

A number of Delawares, Shawnees, and Kickapoos, from Missouri and the borders of Arkansas, have for several years past been engaged in a traffic with the prairie Indians, which has had a tendency to defeat the efforts of the military authorities in checking their depredations upon the citizens of the northern provinces of Mexico. These traders, after procuring from the whites an outfit of such articles as are suited to the wants of the prairie Indians, visit all the different bands, and prosecute a very lucrative business. The goods they carry out consist of a few articles of small value, such as tobacco, paint, knives, calico, wampum, beads, &c., &c., which are of the utmost importance to the Indians, and which, if necessary, they will make great sacrifices to procure; but as they have no commodity for exchange that the traders desire except horses and mules, they must necessarily give these for the goods, and large numbers are annually disposed of in this manner. As I have before mentioned, nearly all these animals are pilfered from the Mexicans; and as the number they traffic away must be replaced by new levies upon their victims, of course all that the traders obtain causes a corresponding increase in the amount of depredations. Should the government of the United States feel disposed to make the prairie Indians annual donations of the same description of articles that the traders now supply them with, (which I am most happy to learn is now contemplated,) upon the express condition that they would continue only so long as they adhered strictly to all the requirements of the agents, it would in a measure obviate the necessity of their making long expeditions into Mexico, and would most undoubtedly have the effect of depreciating the value of the merchandise to such a degree that the traders would no longer find the traffic profitable. The Indians of the plains are accustomed, in their diplomatic intercourse with each other, to exchange presents, and they have no idea of friendship unaccompanied by a substantial token in this form: moreover, they measure the strength of the attachment of their friends by the magnitude of the presents they receive; and I am firmly convinced that a small amount of money annually expended in this way, with a proper and judicious distribution of the presents, would have a very salutary influence in checking the depredations upon the Mexicans. In a talk which I held with a chief of one of the bands of prairie Indians, I stated to him that the President of the United States was their friend, and wished to live in peace with them. He replied that he was much astonished to hear this; for, judging from the few trifling presents I had made his people, he was of opinion that the "Big Captain" held them in but little estimation. Trained up, as the prairie Indians have been from infancy, to regard the occupation of a warrior as the most honorable of all others, and having no permanent abiding-places or local attachments, they can without inconvenience move all their families and worldly effects from one extremity of the buffalo range to the other. With their numerous and hardy horses they travel with great rapidity; and possessing as intimate a knowledge as they do of the localities, it would give them a great advantage over any body of troops who should pursue them into the country. War would not, therefore, be as great a calamity to them as to other tribes who have permanent habitations. Some have supposed that a large body of these Indians could not obtain a sufficient amount of subsistence to enable them to remain together for any great length of time; but their numerous horses and mules, which they often make use of for food when game is scarce, would supply them with subsistence for a long time. It will be necessary to devise some measures to do away with the inveterate prejudices which the Comanches entertain against the habits and customs of the whites, before they will be induced to remain in any fixed abodes or cultivate the soil.

In common with most other Indians, they are very superstitious: they believe in dreams, the wearing of amulets, medicine-bags, &c., and the dedication of offerings to secure the favor of invisible agents; as also in the efficacy of music and dancing for the cure of diseases. They submit with the most imperturbable stoicism and apathy to misfortunes of the most serious character, and, in the presence of strangers, manifest no surprise or curiosity at the exhibition of novelties: yet this apparent indifference is assumed, and they are in reality very inquisitive people. In every village may be seen small structures, consisting of a framework of slight poles, bent into a semi-spherical form, and covered with buffalo-hides. These are called *medicine-lodges*, and are used as vapor-baths. The patient is seated within the lodge, beside several heated stones, upon which water is thrown, producing a dense hot vapor, which brings on a profuse perspiration, while, at the same time, the shamans, or medicine-men, who profess to have the power of communicating with the unseen world, and of propitiating the malevolence of evil spirits, are performing various incantations, accompanied by music on the outside. Such means are resorted to for healing all diseases; and I am also informed that their young men are obliged to undergo a regular course of steam-bathing before they are considered worthy of assuming the responsible duties of warriors. The knowledge they possess of their early history is very vague and limited, and does not extend further back than a few generations. They say that their forefathers lived precisely as they do, and followed the buffalo: that they came from a country towards the setting sun, where they expect to return after death. They acknowledge the existence and power of a great supernatural agent, who directs and controls all things; but this power they conceive to be vested in the sun, which they worship and appeal to on all occasions of moment. They also anticipate a future state of existence similar to the present, and invariably bury with the warrior his hunting and war equipments. Thus far no efforts have ever been made to improve the moral or physical condition of these people; no missionaries have, to my knowledge, ever visited them, and they have no more idea of Christianity than they have of the religion of Mahomet. We find dwelling almost at our doors as barbarous and heathenish a race as exists on the face of the earth; and while our benevolent and philanthropic citizens are making such efforts to ameliorate the condition of savages in other countries, should we not do something for the benefit of these wild men of the prairies? Those dingy noblemen of nature, the original proprietors of all that vast domain included between the shores of the Atlantic and Pacific, have been despoiled, supplanted, and robbed of their just and legitimate heritage, by the avaricious and rapid encroachments of the *white man*. Numerous and powerful nations have already become exterminated by unjustifiable wars that he has waged with them, and by the effects of the vices he has introduced and inculcated; and of those that remain, but few can be found who are not contaminated by the pernicious influences of unprincipled and designing adventurers. It is not at this late day in our power to atone for all the injustice inflicted upon the *red men;* but it seems to me that a wise policy would dictate almost the only recompense it is now in our power to make—that of introducing among them the light of Christianity and the blessings of civilization, with their attendant benefits of agriculture and the arts.

CHAPTER XI.

PACIFIC RAILWAY—IMPRACTICABILITY OF CROSSING THE "LLANO ESTACADO"—ROUTE FROM FORT SMITH TO SANTA FE—RETURN ROUTE FROM DONA ANA—ITS CONNECTIONS WITH THE MISSISSIPPI AND THE PACIFIC.

The very lively interest that has been manifested in a project of such importance as that of uniting the Atlantic with the Pacific by a single span of railroad over the continent of North America, and the prevailing dearth of reliable information regarding a great portion of that vast territory lying west of the Mississippi, induce me to add a few remarks upon this subject, which I trust will not be wholly devoid of interest or utility at this particular period.

Whether this road should be national, and its administration under the direction of the general government, or whether it should be intrusted to individual States or corporate companies, are questions the discussion of which it does not become me to attempt, and upon which I shall not presume to hazard an opinion. I propose, in what I have to say, merely to give a brief detail of such facts connected with this subject as are suggested after an examination of a district of country over which it may be found desirable to construct the road.

Although the appropriation made by the last Congress for preliminary surveys, indicates a disposition on the part of our national legislature to give aid in the initiatory steps, and although great benefits would undoubtedly result from bringing our distant possessions in the west into closer proximity with the eastern States, by a means of transit much more expeditious than any which nature offers, thereby facilitating the transmission of troops and munitions of war, the value of the project, in a commercial aspect, appears to be of sufficient magnitude to denote a reasonable guarantee for its speedy execution.

The importance, and indeed the necessity, of this road, are very generally admitted. It is the will of a people controlling a great share of the commerce of the world that it should be made; and possessing, as they do, ample pecuniary resources, and stimulated by the ambitious but laudable prospect of turning and monopolizing the channel of Asiatic trade, with the almost certain anticipation of profit, it is easy to predict the result. The financial demonstration recently made in New York city, whereby an amount of stock almost sufficient to carry out the enterprise was subscribed in one day, is eminently significant of the fact, and affords substantial evidence of the confidence of capitalists in the feasibility of the scheme, and an abundant pledge for its early and successful accomplishment. That the road will be constructed, but few at this time entertain a doubt; the only question that remains to be determined is, where is the best and most advantageous route?

The several exploring parties that have been sent across the continent in different latitudes, will undoubtedly place the department in possession of all the information required concerning the country over which the limited amount of the appropriation, and time, enabled them to pass; but as a large portion of the district over which I have travelled will not come within the scope of their reconnoissances, my remarks may serve to throw some light upon the subject, which cannot be obtained from other sources, and thus add to the general stock of information so earnestly sought after at this particular period.

The district of country to which my attention has been directed is embraced within the 32d and 36th parallels of latitude, and the 95th and 107th meridians of longitude; and is bounded upon the north by the Canadian river, and upon the west by the Rio Grande. A great portion of this vast domain, containing nearly thirty-two thousand square miles, was previous to 1849 almost wholly unknown, except to the native occupants.

One of the most prominent features which strikes the eye of the beholder on an examination of this section, is the very remarkable uniformity of its surface, and the almost total absence of those abrupt and rugged primitive mountain ranges which in many other parts of our country offer such formidable obstacles to the passage of railways. But few mountains are seen thoughout this region, and those few are so little elevated that they present but trifling obstructions when compared with many that are found in the eastern States. This section is, however, traversed throughout, nearly its whole length, by the lofty plateau of the "Llano estacado," which, as will be observed upon the map, stretches out from the 32d to the 36th parallel of latitude, and is, in places, two hundred miles wide, without a tree or running stream throughout its entire surface, and presents, in my judgment, an impassable barrier to a wagon road; and I am fully impressed with the belief that a route crossing this desert anywhere between the 33d parallel of latitude and its northern limits will never be selected for a Pacific railway, or, indeed, a road of any description. South of this parallel the plain becomes less elevated above the adjacent country, and finally merges into the lands bordering the Pecos and the head branches of the Colorado.

If it be impracticable to construct and find the material for sustaining a railway across this desert, the question arises whether a feasible, route can be found near the northern or southern borders of it.

The road which was made under my supervision from Fort Smith Arkansas, to Santa Fé, New Mexico, in 1849, (with one exception, where it crosses a spur, which can easily be turned,) skirts the base of the northern border of this plain; and so far as the topography of the country is concerned, I believe that a railroad can be made over it with great facility, as the general surface is smooth, and intersected by no impassable mountains or deep valleys.

On departing from Fort Smith, this road traverses a gently undulating district, sustaining a heavy growth of excellent timber, but occasionally interspersed with prairie lands, affording luxuriant grass for eight months in the year, and intersected with numerous small streams flowing over a highly productive soil, thus embracing the elements of a rich and beautiful pastoral and agricultural locality. This character continues for one hundred and eighty miles, to near the 99th meridian of longitude, where the road emerges from the woodlands and enters the great plains, where but little timber is seen except directly along the borders of the water-courses. The soil soon becomes thin and sandy, and, owing to the periodical droughts of the summer season, would require artificial irrigation to make it available for cultivation.

Soon after leaving the woodlands the road takes a ridge which divides the Canadian from the Washita river, and continues upon it to near the sources of the latter stream, a distance of nearly three hundred miles. This ridge lies in a very direct course for Santa Fé, is firm and smooth, and makes one of the best natural roads I have ever travelled over. The ground upon each side is cut up into a succession of deep and precipitous gullies, which have been washed out by the continued action of water in such a manner as to render any other route in the vicinity, but the one directly upon the crest of the "divide," almost impassable.

From the head of the Washita the road continues near the valley of the Canadian for a hundred miles further, occasionally crossing small tributaries which furnish the traveller with water at convenient distances; it then bears to the left, and passes over the elevated lands bordering the Pecos river, skirting the base of the mountains along that stream until it arrives at a place called "Laguna Colorado," a small lake of muddy water, where the road forks, one branch leading to Santa Fé over a road forty miles in length, and the other to Albuquerque, (the point where the route through what is called "Walker's Pass" is said to leave the Rio Grande,) a distance of only twenty miles.

The distance from Fort Smith to Santa Fé, as measured with the chain, is eight hundred and twenty miles.*

The line of this road continued east from Fort Smith would intersect the Mississippi river in the vicinity of Memphis, Tennessee, and would pass through the country bordering the Arkansas river, which cannot be surpassed for fertility, as the bountiful crops of cotton, corn, and other products grown by the planters, abundantly evince.

The route of my return from New Mexico in 1849, which has been travelled by California emigrants every year since that time, leaves the Rio Grande at a point called Doña Ana, three hundred miles below Santa Fé.

On leaving this place, at an elevation of about four thousand feet above the sea, the road for three hundred miles traverses an arid prairie region, where but little wood is found except upon three ranges of mountains which stretch out to the north, but do not materially obstruct the passage of the road. They are covered for the most part with pine timber, and abound in springs of wholesome water, making it imperative upon the traveller to pass near them. Upon the route marked down, the defiles have but little elevation above the general surface, and, with the exception of a few miles of broken ground near the "Peak of Gaudalupe," the ascents and descents to all the undulations are gradual and easy. At the southern extremity of the Gaudalupe mountains the summit level of the country between the Rio Grande and the Pecos is attained, and from this point the surface declines to the borders of the latter stream by a gradation almost imperceptible. Crossing the Pecos, the road ascends by a grade of about five feet per mile for twenty-five miles, and the traveller here finds himself upon the broad plain of the "Llano estacado," which at this point divides the waters of the Rio Grande from those of the Colorado. The road crosses the southern spur of this plain, where it is seventy miles broad, and as firm and smooth as the best McAdamized road. Thence it crosses the head branches of the Colorado and the main Brazos, and leads off to a ridge which terminates near Fulton, Arkansas, upon the navigable waters of Red river. By leaving this ridge and crossing Red river at Preston, a good

---

*The barometrical altitude of Albuquerque above tide-water is about 5,130 feet, and of Fort Smith about 600 feet; making the difference in altitude, or total declination eastward between the two points, 4,530 feet, or an average grade of a little over 5⅝ feet to the mile.

road is found to Fort Smith, upon the navigable portion of the Arkansas, which would be in a very direct course for St. Louis, and traverse one of the most productive sections of the United States.

The entire distance from Doña Ana to Fulton is about eight hundred and fifty miles, and to Fort Smith nine hundred and four miles. The road from El Paso connects, at the Sierra Waco, with the one described, and is thirty miles shorter.

Doña Ana being elevated four thousand feet above the tide-water level, and Fulton and Fort Smith six hundred and sixty and six hundred respectively, gives an average grade of less than four feet to the mile over either road. These results, of course, can only be regarded as approximate estimates, which will be increased upon the undulatory portions of the routes. The surface of the country, however, has a remarkably uniform dip to the east and south throughout nearly its whole extent, and is perhaps better adapted by nature to the reception of a railroad than almost any other which can be found.

A glance at a map of the country will show that Red river, from the point of its efflux upon the Delta of the Mississippi to Fulton, has a northerly bearing; that here it makes a sudden deflection of almost a right-angle to the west, and maintains this course to its origin in the "Llano estacado."

The road alluded to, immediately after leaving Fulton, leads to an elevated ridge, dividing the waters that flow into Red river from those of the Sulphur and Trinity, and continues upon it, with but few deviations from the direct course for El Paso and Doña Ana, to near the Brazos river, a distance of three hundred and twenty miles. This portion of the route has its locality in a country of surpassing beauty and fertility, and possesses all the requisites for attracting and sustaining a dense farming population. It is diversified with prairies and woodlands, affording a great variety of excellent timber, and is bountifully watered with numerous spring-brooks, which flow off upon either side of the ridge into the streams before mentioned. The crest of the ridge is exceedingly smooth and level, and is altogether the best natural or artificial road I have ever travelled over for the same distance.

After leaving this ridge the road crosses the Brazos near very extensive fields of bituminous coal, (the only locality of this mineral, so far as my knowledge extends, that has been discovered within two hundred miles,) which burns readily with a clear flame, is made use of for fuel at Fort Belknap, and is very superior in quality.

From the Brazos the road skirts small affluents of that stream and the Colorado for two hundred miles, through a country more undulating than that east of the Brazos; but no mountains are met with, or elevated hills, which cannot be avoided by short detours.

Here and there prairies present themselves, but this section is for the most part covered with a growth of trees called mezquite, which stand at such intervals that they present much the appearance of an immense peach orchard. They are from five to ten inches in diameter, their stocks about ten feet in length, and for their durable properties are admirably adapted for railway ties, and would furnish an inexhaustible amount of the very best fuel. The soil upon this section is principally a red argillaceous loam, similar in appearance to that in the Red river bottoms, which is so highly productive, and extends to near the 102d degree of longitude, or about three degrees further west than the arable soil upon some of the more northerly routes.

As this route is included within the 32d and 34th parallels of latitude, it would never be obstructed by snow, as it seldom falls more than two or three inches in depth, and only remains upon the ground a few hours at a time.

The whole surface of the country, from Red river to the Rio Grande, is covered with a dense coating of the most nutritious grass, which remains green for nine months in the year, and enables cattle to subsist the entire winter without any other forage.

It will be observed that the route here spoken of skirts the head-waters of the rivers flowing towards the Gulf of Mexico, for several hundred miles after leaving Red river, and that a road cannot be made much further to the north without impinging upon the "Llano estacado." From what I have seen of the country south of this, I have no doubt but that a road could be made in almost any direction, but would be attended with much greater cost than upon the one I have attempted to describe, for the reason that the surface of the country along this route is much more level.

After passing the Brazos river, the road, as I have before observed, runs near the sources of the streams, where the valleys are broad and but little depressed below the general surface; whereas I have remarked that in descending some of these streams, the longitudinal and lateral valleys become deep and abrupt, and where (as would be the case with a Pacific railway) it became necessary to cross these undulations transversly, a greater expenditure of labor would be involved in grading than upon the other route. There would also be many more large streams to bridge; indeed, upon the route I have recommended, there are but two streams (the Brazos and Pecos) of greater width than forty feet, over the entire distance of eight hundred and fifty miles, between Red river and the Rio Grande.

As Fulton, El Paso, and San Diego, in California, are nearly in the same direct line, and one which intersects the longitudinal axis of the continent at right-angles, a road connecting these points would form the shortest line of communication to the Pacific in this latitude, and would pass near the valley of the Gila, or its vicinity.

The direct line of this road prolonged eastwardly from Fulton would pass through Arkansas, and intersect the Mississippi river a little below Napoleon, opposite the State of Mississippi, and would traverse a section which presents no serious impediment to the passage of a railroad.

This route was surveyed in 1851 by Mr. Sidell, (civil engineer,) under the direction of the Topographical Bureau, and resulted in perfectly establishing the feasibility of the route, and the determination of the fact that the most elevated ground between Lake Providence and Fulton (a distance of two hundred miles) is but one hundred and sixty feet above the flood water-table of the Mississippi, and only one hundred feet above that of Red river.

The terminus of the eastern section of this route upon the Del Norte, could be resumed upon the west bank of that stream; and if the practicability of constructing the road down the valley of the Gila can be established, it would give a continuous line to the Colorado river.

Although our knowledge of the country west of the Rio del Norte is for the most part confined to a few traces that have been pursued by travellers making their way to the Pacific; yet it is believed that sufficient reliable data may be deduced from competent authorities to warrant the expectation of finding a route with admissible grades, as far, at least, as the Colorado.

Before reaching the waters that flow into the Pacific, it becomes necessary upon this route, as upon all others in our territory, to surmount the Rocky Mountain chain. The elevation of the crest of this great continental vertebral column varies from five to seventeen thousand feet above the tide-level of the ocean, but has a declension towards its southern extremity, which greatly favors the project in question. The barometrical measurements which have been made, place "Long's Peak" in latitude 40° 36′ at the maximum, and the culmination of a pass or defile in near latitude 32° at the minimum altitude.

The elevation of the eastern base of the mountains in latitude 42° is the same as the summit of the range in latitude 32°. The elevation of other passes that have been examined, vary from seven to eight thousand feet above tide.

If, so far as the economy of railway transportation is concerned, the attainment, with the locomotive, of twenty feet in altitude, is equivalent to the transit of a mile upon a horizontal plane, we would have (other conditions being equal) a difference of one hundred miles in horizontal distance in favor of the route under consideration, over one which should pass the mountains at an elevation of seven thousand feet.

The difference of elevation of the Rio del Norte in the vicinity of Doña Ana, and the crest of the mountains in latitude 32°, being about one thousand feet, and the distance between the two positions about one hundred miles, gives an average grade (which is said to be very uniform) of ten feet per mile in ascending the eastern slope of the mountains. From the summit to the mouth of the Gila, a distance of three hundred and eighty miles, the difference in altitude, barometrically determined, is four thousand seven hundred and forty-six feet, which (supposing the slope to be uniform) admits of a gradient of about twelve and a half feet to the mile in descending the Pacific side of the mountains.

The road upon this side would be much more circuitous in its course than upon the other; the grades will be increased upon the undulatory portions of the country, and some difficulty may be anticipated in passing the great cañon of the Gila, provided the road is confined exclusively to the limits of our own territory; but a gentleman of scientific attainments, who has examined this route carefully, is of the opinion that no greater impediments will be met with upon the Gila than are found upon the Hudson river road. From the Colorado to the Pacific (unless some other pass is discovered) the road must penetrate the "Sierra Nevada" chain, through what is called "Warner's Pass."

The summit of this defile is situated north of the general course of the road, and the approaches to it, upon both sides, are somewhat tortuous. It is about one hundred miles distant from the confluence of the Gila with the Colorado, and about eighty miles from San Diego on the Pacific.

The elevation of the Pass above tide-water being three thousand and thirteen feet, and that of the Colorado, at the mouth of the Gila, two hundred and fifty-four feet, we would have an average gradation of twenty-seven and a half feet per mile in the ascent of the eastern slope, and a descent from the Pass of thirty-seven and a half feet to the mile in reaching San Diego.

Should it be found desirable, on arriving at the mouth of the Gila, to turn the course of the road, and run it down the valley of the Colorado to the head of the Gulf of California, I am informed by persons who have examined this section that the surface is free from obstruc-

ADAIR BAY.

tions, and the distance to Adair bay (where four fathoms of water may be relied upon at ebb-tide) is about eighty miles.

    I am, sir,
        Very respectfully,
            Your obedient servant,
                RANDOLPH B. MARCY,
                  Captain 5th Infantry, U. S. Army.

APPENDIX A.

METEOROLOGICAL OBSERVATIONS.

120   APPENDIX A.—METEOROLOGICAL OBSERVATIONS.

Table of Meteorological Observations kept during the expedition for the exploration of the Red river and its sources, under charge of Captain R. B. Marcy, 5th U. S. infantry.

Date.	Hour.	Barometer.	Thermometer.		Wind.		Weather.	Remarks.
			Attached.	Detached.	Direction.	Intensity.		
1852.			°	°				
April 12	8.30 p. m.	29.5					Cold and fair	At Fort Smith water froze at night.
27	Sunrise			65			Cold and fair	
28	Sunrise			32			Clear	
May 1	3 p. m.	28.25		52			Cloudy	Fort Belknap.
2							do	Do. do.
3							do	In camp.
4	4 p. m.	28.4		88	S. E.	Moderate	do	Do.
4	9 p. m.	28.473		76	S. E.	Violent	Hazy	Rained in the afternoon.
5	3 p. m.			77			Cloudy	Rained violently in the morning.
6								
7	3 p. m.	28.8	80	80	S. E.		Clear	A few slight clouds; sun shining.
7	7.30 p. m.	28.7	73	72½	S. E.		Cloudy	Threatens rain; lightning in N. W.
7	9 p. m.	28.75	71	70	S. E.		do	Raining slightly.
8	5 a. m.	28.71	69½	68½	S. E. by S.		do	Rained during the night.
8	10 a. m.	28.753	80	80	S. E.	Moderate	do	
8	2 p. m.	28.72	83	84½	S. E.	do	do	
8	3.30 p. m.	28.654	83	81	S. E.	do	do	
8	8.30 p. m.	28.55	74	72	S. E.	Strong	do	Lightning in N. W.
9	6.30 a. m.	28.542	69½	68	E.	Light	do	Rained during the night.
9	1 p. m.	28.6	86	86	S. E.	Moderate		Sun shining at intervals.

APPENDIX A.—METEOROLOGICAL OBSERVATIONS. 121

Day	Time	Bar.		Temp	Wind	Force	Clouds	Remarks
9	3.30 p. m.	28.55	82	79½	S. E.	Strong	Cloudy	Sun has been shining for a time; clouds now coming up; thunder in the west.
9	7 p. m.	28.5	76	76	S. E.	do	do	Vivid heat-lightning in N. W.
9	9 p. m.	28.5	72¾	70	S. E. by S.	do	Clear	
10	5 a. m.	28.552	69½	68	S. E.	Slight	Cloudy	
10	9.30 a. m.	28.578	78	78	S. E.	Strong	do	Light clouds; sun shining at intervals.
10	12 m.	28.503	84	86	S. E.	Moderate	do	
10	4.30 p. m.	28.49	88	87	S. E.	Strong	Clear	Light clouds around the horizon.
10	7 p. m.	28.49	76	76	S. E.	Slight	do	Do. do.
10	8.30 p. m.	28.5	73	69	S. E.	Strong	do	
11	4 a. m.	28.55	67	64	S. E.	Very light	do	
11	6.30 a. m.	28.7	68	68	S. E.	do	do	In Red River Bottom.
11	10.30 a. m.	28.725	88	89	S. E.	do	do	Heat-lightning in the north.
11	5 p. m.	28.65	85	82	S. E.	do	do	Light floating clouds; patches of blue sky visible; very beautiful; morning red.
11	8.30 p. m.	28.67	74	71	S. E.	do	do	
12	5 a. m.	28.67	70	68¼	S. E.	do	Cldy	
12	6 a. m.	28.67	70	68½	S. E.	do	do	Rain-storm from N.W. about 7 a. m.
12	4 p. m.	28.55	80	77	S. E.	do	do	Raining; thunder and lightning in the west.
12	7.30 p. m.	28.54	74½	73	S. E. by S.	do	do	
12	9 p. m.	28.55	73½	72	S. E.	do	Part cldy	Light clouds; many stars visible; lying in N. W.
13	6 a. m.	28.54	71	69	N. W.	do	Cldy	It ran; thunder and lightning.
13	7 a. m.	28.4	67	65	N. W.	Light	do	Rain has ceed; clearing off.
13	9 a. m.	28.5	71½	69	W. N. W.	do	Clear	
13	2 p. m.	28.5	90	85	N. W.	Moderate	do	
13	9 p. m.	28.6	67¼	64	N. W.	do	do	
14	5 a. m.	28.65	57	52½	N. W.	Light	do	Rather heavy dew.
14	6.30 a. m.	28.65	62	57¼	W. N. W.	do	do	

122 APPENDIX A.—METEOROLOGICAL OBSERVATIONS.

METEOROLOGICAL OBSERVATIONS—Continued.

Date.	Hour.	Barometer.	Thermometer.		Wind.		Weather.	Remarks.
			Attached.	Detached.	Direction.	Intensity.		
1852.		″	°	°				
May 14	9 a. m.	28.768	76½	76	S. W.	M'rate	Clear	
14	12 m.	28.75	82	82½	S.	do.	..do.	
14	2 p. m.	28.76	83	84½	S.	..do.	..do.	
14	10 p. m.	28.636	70½	69	S. E.	Light.	Misty.	Lightning in the north.
15	5 a. m.	28.6	68½	66	S. E.	..do.	Clear.	
15	8.30 a. m.	28.6	75½	74	S. E.	Moderate.	Hdy.	
15	3 p. m.	28.55	82	81	S. E.	..do.	..do.	Lightning in the N. and N. W.
15	9.30 p. m.	28.55	74	72	N.	Very light.	..do.	
16	5 a. m.	28.65	68	66	N.	do.	..do.	Ring around the sun.
16	4.30 p. m.	28.856	68½	64	N. N. E.	Moderate.	..do.	
16	7.30 p. m.	28.85	59½	57½	N. N. W.	..do.	..do.	A few stars dimly visible.
16	9.45 p. m.	28.85	56½	52	N.	Light.	..do.	
17	5 a. m.	28.75	46½	44½	N.	..do.	..do.	
17	5 p. m.	28.628	67	61	N.	..do.	Clear.	
17	7.45 p. m.	28.588	55	51	N.	..do.	..do.	
17	9.30 p. m.	28.62	50	44½	N.	do.	..do.	
18	4.30 a. m.	28.58	42	40	N. E.	Very light.	..do.	
18	3.45 p. m.	28.424	81	79	S.	do.	..do.	
18	6 p. m.	28.5	75	73	S. E.	Light.	Half cloudy.	Heat-lightning in the S.
18	8.45 p. m.	28.52	72	70	S. E.	Moderate.	do.	Rain; thunder and lightning; wind increasing.

APPENDIX A.—METEOROLOGICAL OBSERVATIONS. 123

Day	Time				Wind		Sky	Remarks
19	4 a. m.	28.45	66	61	E.	Strong	Cloudy	Has been raining violently, with thunder and lightning, for eight hours.
19	5 a. m.	28.5	63	60	E. S. E.	do.	do.	Rain, with thunder and lightning.
19	6 a. m.	28.57	61	59	N. N. E.	do.	do.	Do. do. do.
19	8.30 a. m.	28.65	61½	60	N. N. E.	do.	do.	Rain ceased; a little blue sky visible.
19	12 m.	28.6	63¾	61	E. E. N.	do.	do.	Rained on the 20th and 21st in the morning.
21	6 p. m.			75	N. E.	Moderate.	Clear.	
21	9.45 p. m.			66	N. E.	do.	Cloudy	Lightning in S. at 7½ p. m.; rain storm arose from S. at 8½ p. m.
22	4.45 a. m.			62	N. W.	do.	do.	No indication of rain; no lightning.
22	6.15 p. m.			71½	N. E.	do.	do.	
22	9 p. m.			65	N. E.	do.	do.	
23	12.45 a. m.			64	N. E.	do.	do.	
23	6 p. m.			60½	N. E.	do.	do.	
23	7 p. m.		70½	58	N. E.	do.	do.	Raining moderately.
23	2.30 p. m.	28.5		65	N. E.	do.	do.	Severe storm (u&r, lightning, and rain) th&n 10 and 1½ p. m.
23	6 p. m.	28.4	71¼	65¼	E. N. E.	do.	do.	Light &lds.
23	7 p. m.	28.356	70	62	E. N. E.	do.	Partly cloudy.	A few light clouds near the &m.
23	9.15 p. m.	28.415	68½	62	E. N. E.	Light.	Clear.	&us clouds &d in reti u-lated form; no &nt rain storm.
24	4.30 a. m.	28.352	67	62	E. S. E.	S&g.	Cloudy	&g.
24	6 a. m.	28.35	68	62	N.	&te.	do.	Do.
24	8.30 a. m.	28.26	68	62	N.	Strong.	do.	&g; &us &.
24	10.30 a. m.	28.25	70	65	N. E.	Light.	Partly &dy.	Sun shining; clouds &u&us and stratus.
24	11.30 a. m.	28.251	73	67	S. E.	Moderate.	Partly &dy.	
24	12.30 p. m.	28.34	83¼	80	S.	do.	&dy.	N t &his clouds; blue &ible.
24	2 p. m.	28.335	84¼	74	S.	Light.	do.	Nimbus &.
24	4.30 p. m.	.283	81	74¼	S.	do.	do.	Do. do.
24	5.30 p. m.	28.26	76¾	72	S.	do.	do.	Do. do.

METEOROLOGICAL OBSERVATIONS—Continued.

Date.	Hour.	Barometer.	Thermometer.		Wind.		Weather.	Remarks.
			Attached.	Detached.	Direction.	Intensity.		
1852. May 24	7.30 p. m..	″	° 74	° 65	S.	Light....	Clear......	Cl'ds east; stratus north (horizon.)
25	1.30 a. m..	28.28	64	60	S.	..do.....	Cloudy.....	Nimbus and stratus clouds; dew at 8½; it rains \ on 11 and 1 a. m.
25	6.30 a. m..	28.325.	68½	65	S. E.	Moderate..	..do......	Foggy sky.
25	8 a. m.....	28.33	70	64	S.	Light.....	..do......	Cl'ds and rarius cl'ouds; blue sky; has l'en raining.
25	9.30 a. m..	28.33	73½	68	S.	..do.....	..do......	Sun shining; rainy; cumulus cl'ds.
25	2.45 p. m..	28.375	93	86	S.	..do.....	..do......	Stratus and nimbus cl'ds; raining badly.
25	5.30 p. m..	28.315	80	74	W. N. W.	..do.....	..do......	Stratus and n's cl'ds around us; heat-lightning in E.
25	7.30 p. m..	28.3	76	67½	W. N. W.	..do.....	..do......	Sli'gt dw'; ling in the east.
25	8.30 p. m..	28.3	74½	65	W. N. W.	..do.....	..do......	Nimbus and n's cl'ds.
25	11.30 p. m.	28.3	69	64	N. W.	..do.....	Cloudy.....	
26	6.15 a. m..	28.3	67½	62	N. W.	..do.....	..do......	
26	11.30 a. m.	28.3	79	76	N. W.	..do.....	..do......	Rained at 11 p. m.
26	1.30 p. m..	28.35	78¾	76	N. W.	..do.....	Clear.....	A few cumulus clouds near the horizon.
26	3 p. m.....	28.38	89	82	N. W.	..do.....	Cloudy.....	Raining slightly; sun shining; nimbus and cumulus clouds.

APPENDIX A.—METEOROLOGICAL OBSERVATIONS.

Date	Time	Bar.			Wind	Force	Sky	Remarks
26	5 p. m.	28.35	87	76	N. W.	do	Clear	A few clouds floating
26	6 p. m.	28.375	87	72	N.	do	do	Shower from N.; sun shining; perfect rainbow; stratus, nimbus, and cumulus clouds.
26	8.30 p. m.	28.33	73½	64	N.	do	do	Dew commences falling.
26	11.45 p. m.	28.325	69	60	N. W.	do	do	Heavy dew.
27	5.30 a. m.	28.31	63	61½	N. W.	Very light	do	Dew; cumulus clouds in the east.
27	6.15 a. m.	28.33	68	66	N.	do	do	
27	Midnight	28.3	72	64	N.	Calm	do	Dew.
28	5.30 a. m.	28.3	6½	58	N.	do	do	
28	7 a. m.	28.25	6½	64	N.	do	do	
28	9 a. m.	28.35	84	76	N.	do	do	
28	6 p. m.	28.25	84	84	W. by S.	Very lgt	do	No dew; beautifully clear night.
28	7.30 p. m.	28.25	79	69	N. W.	do	do	Dew.
28	11 p. m.	28.225	69	64	S. W.	do	do	Dew.
29	3.30 a. m.	28.2	64	61	S. W.	do	do	Dew.
29	9 p. m.	28.25			S. W.	do	do	Dew; beautiful night.
30	9.30 p. m.	28.25	78	70	S. W.	Calm	do	
31	3.30 a. m.	28.225	68		S. W.	d	do	
31	10.15 p. m.	28.250	72	66	S. E.	dry lgt	do	Slight dew.
June 1	2.30 a. m.	28.2	70	63	S. W.	Moderate	do	Heavy dew.
1	2 p. m.	28.25	97	88	S. W.	lgt	do	Stratus clouds.
1	7.30 p. m.	28.15	82	76	S. E.	Strong	do	No dew.
1	9.45 p. m.	28.075	77¾	77¾	S.	do	do	Do.
2	2.45 a. m.	28.06	74	69	S.	do	do	Do.
2	11.15 a. m.	28.025	91	84	S.	do	do	Do.
2	3.30 p. m.	28.025	100	87¾	S.	do	do	Do.
2	4.30 p m.	27.95	102	88	S.	do	do	Stratus clouds in the west; richly colored sunset.
2	7 p. m.		87½	9	S. E.	Moderate	do	No dew; lightning in N. W.
2	10.15 p. m.	27.95	77	74	S. E.	do	do	Stratus clouds around horizon; no dew.
3	2.15 a. m.	27.9	75	70	S. E.	do	do	

126   APPENDIX A.—METEOROLOGICAL OBSERVATIONS.

METEOROLOGICAL OBSERVATIONS—Continued.

Date.	Hour.	Barometer.	Thermometer.		Wind.		Weather.	Remarks.
			Attached.	Detached.	Direction.	Intensity.		
1852.		"	°	°				
June 3	10.15 a. m.	27.925	98	89	S. E.	Calm	Cloudy	Stratus cl t d; rained v in 6h. ad 6h. 30m. a. m.; morning d.
3	12 m.	28	99	86¾	N. E.	Strong	..do..	Stratus lds.
3	2.15 p. m.	28.525	98	86¼	N.	Violent	..do..	lds louds; an big coa- sionally.
3	7.30 p. m.	28.55	80	70	N.	Strong	Clear	Stratus nd lds cl ds nr din.
3	9 p. m.	28.53	73½	70	N.	Violent	..do..	Stratus nd lds lds ar din; lightning in ush.
4	2.30 a. m.	28.55	64	60	N.	..do..	Hazy	Ring around he rm; no vd.
4	1.30 p. m.	28.6	82¼	76	N.	Strong	Clear	Stratus clouds; ng red at sun- se.
4	5 p. m.	28.57	81	74	N.	..do..	..do..	Stratus lds.
4	7 p. m.	28.53	72½	67¾	N.	Light	..do..	A few l gt stratus ds around the ht set.
4	8.30 p. m.	28.525	66¾	62¾	N.	..do..	..do..	No vd.
4	9.30 p. m.	28.5	61	58	N.	..do..	..do..	No vd.
4	11.30 p. m.	28.475	55	52	N. W.	..do..	..do..	Slight dw; stratus clouds in S. W.
5	1.30 a. m.	28.46	52	49½	N. W.	..do..	..do..	Dew; morning red at rse.
5	12 m.	28.45	80	77	N. W.	..do..	Cloudy	
5	4.30 p. m.	28.39	80	76	N. W.	Very light.	..do..	Stratus clouds.
5	5.30 p. m.	28.35	77½	74½	S. W.	..do..	Clear	

APPENDIX A.—METEOROLOGICAL OBSERVATIONS.

5	7 p. m.	28.3125	72	65	S. W.	Light.	...do...	Stratus ds in the horizon to the west.	
5	9.30 p. m.	28.27	60½	59	S.	Very lgt.	...do...	No wd	
6	2.30 a. m.	28.2	55	54	S.	Ge.	Cloudy.	Moderate wd; morning red.	
6	12 m.	28.19	76	73	N.W.	Light.	...do...	Nl us d; ing.	
6	3 p. m.	28.18	81	76	N.W.	...do...	...do...	Raining.	
6	4½	28.15	73	71	N.W.	...do...	...do...	Nl us nd stratus ld;	
6	7 p. m.	28.175	71	64	N.	...do...	...do...	Stratus d; hs bn raining tlly at intervals. rain continued 4 hours.	
6	p.m.	28.225	69	64	N.		...do...	Nimbus clouds; morning red.	
7	am.	28.2	61½	58½	N.	d...	...do...	Cumulus clouds.	
7	12 m.	28.25	82	83	N.		...do...	A few cumulus clouds near the horizon.	
7	p.m.	28.275	83	79	N.	dry st'ng.	Clear.		
7	4½	28.28	76	72	N. E.	Strong.	...do...	A few stratus clouds near the horizon.	
7	7 p. m.	28.29	72	67	N. E.	Moderate.	Cloudy.	Stratus clouds.	
7	8.30 p. m.	28.28	67	62	N. E.	Light.	...do...	No dew.	
8	2.30 a. m.	28.21	53	46	N. E.	...do...	Clr.	Slight dew; morning red.	
8	12 m.	28.12	82	75	S.	...do...	...do...	Stratus clouds near the horizon.	
8	2.30 p. m.	28.06	88	80	S.	...do...	...do...	Light stratus and cumulus clouds.	
8	7.30 p. m.	27.92	75	65¾	S.	...do...	...do...		
8	8.30 p. m.	27.89	67	62	S.	...do...	...do...	No dew.	
8	9.45 p. m.	27.875	60	56	S.	...do...	...do...	No dew.	
9	2.30 a. m.	27.825	54½	49	S.	...do...	...do...	Slight dew; morning red.	
9	11.15 a. m.	28.03	77	83	N. E.	Moderate.	...do...	A little hazy.	
9	3 p. m.	28.13	92	81½	N. E.	Strong.	...do...		
9	5.30 p. m.	28.11	86	76½	N. E.	d			
9	7.30 p. m.	28.06	72	66	N. E.	...do...	...do...	A few stratus clouds in the horizon.	
9	8.30 p. m.	28.06	69	63	N. E.	...do...	...do...	A few stratus clouds in the horizon; no dew.	
9	10.30 p. m.	28.075	65	60½	N. E.	...do...	...do...	No dew.	

128 APPENDIX A.—METEOROLOGICAL OBSERVATIONS.

METEOROLOGICAL OBSERVATIONS—*Continued.*

Date	Hour.	Barometer.	Thermometer.		Wind.		Weather.	Remarks.
			Attached.	Detached.	Direction.	Intensity.		
1852.		"	°	°				
June 10	3 a. m.	28.1	59	59	E.	Moderate.	Cloudy	No dew; morning red.
10	11.30 a. m.	28.02	80	78	S. E.	Strong	Clear	A few floating clouds.
10	4 p. m.	27.95	90½	83	S. E.	do	1 dy	
10	6 p. m.	27.95	89¼	84	S. E.	do	d	Stratus clouds.
10	7.30 p. m.	27.9	85	79½	S. E.	do	Cl dy	No dew; nimbus and stratus clouds.
10	9.30 p. m.	27.92	78	74	S.	Moderate.	d	No dew; morning sky red to-day.
11	2.45 a. m.	27.85	66	60	S.	Light.	d	Stratus clouds.
11	12 m.	27.75	96	93	S.	Violent.	d	
11	3 p. m.	27.75	100	92	S.	do	Cl ar	No dew.
11	6 p. m.	27.725	93½	80	S.	Stro	d	No dew.
11	8 p. m.	27.72	82	78½	S.	do	d	No dew; stratus and nimbus clouds; morning sky red.
11	9.30 p. m.	27.71	77	76	S.	do	d	
12	3 a. m.	27.65	73	69½	S.	do	Cloudy	
12	11 a. m.	27.63	89	85	S.	do	Clear	Hazy.
12	2.30 p. m.	27.59	94¼	86½	S.	do	do	No dew.
12	5.30 p. m.	27.53	86	82	S.	do	do	No dew.
12	7.30 p. m.	27.525	81	77	S.	do	do	
12	9.30 p. m.	27.66	77½	72	S.	do	do	
13	5.30 a. m.	27.66	66	63	S.	Moderate.	do	Cumulus clouds floating near the horizon.
13	2.30 p. m.	27.67	92	90	S.	Strong	do	

APPENDIX A.—METEOROLOGICAL OBSERVATIONS.   129

13	4.30 p. m	27.64	88½	83	S.	do	do	this cl ds br he obn.
13	6.30 p. m	27.64	86	80½	S.	do	do	ids ds br the
13	7.30 p. m	27.65	82	77	S.	do	do	ids cl ds br he li.
13	10 p. m.	27.675	76½	72	S.	Moderate	do	dv.
14	2.30 a. m	27.66	67½	66	S.	Strong	do	ig sky red to-day. ds; uje dy ad
14	2.30 p. m	27.63	96	84	S.		do	ids pd every ge of rain at 7½ a. n.
14	6 p. m	27.58	86	79	S.	do	do	ids ds.
14	7.30 p. m	27.575	77	72	S.	Light	do	ids, i lis, & stratus clouds in he uh.
14	9.30 p. m	27.58	72½	68	S.	do	do	No ds.
15	3.30 a. m	27.53	71	66	S.	Strong	Cloudy	ids cl d; gt de at
15	12 m	27.525	92	86	S.	Moderate	Clear	intervals, between 5 a. n. ad 8½ a. m.
15	3.30 p. m	27.49	86	78	S. W. by W.	Strong	Cloudy	Violent rain has just commenced; continued but for 15 minutes.
15	9 p. m.	27.45	75	67	S. W. by W.	Very light	do	No dew.
16	3 a. m	27.4	70	65	S. W.	do	do	Rained at intervals between 5 and 8 a. m.
16	1.30 p. m	27.4	82	84	S. W.	do	do	ids lds.
16	6 p. m	27.38	77	72	S. E.	Light	do	Nimbus clouds.
16	7.30 p. m	27.4	64	68	S.	do	do	Raining slightly.
16	10.30 p. m.	27.4	62	66	S.	do	do	ied all njt.
17	4.30 a. m	27.4	70	64	S.	do	do	ids nd i lis cl ds; slight rain at 10 a. n.
17	12 m	27.5	85	83	S.	Moderate	do	
17	6 p. m	27.5	76	70	S.	do	do	ids nd i lis ds; rain at 3 p. m.

9

METEOROLOGICAL OBSERVATIONS—Continued.

Date.	Hour.	Barometer.	Thermometer.		Wind.		Weather.	Remarks.
			Attached.	Detached.	Direction.	Intensity.		
1852.		″	°	°				
June 18	4.30 a. m	27.49	70	66	W.	Moderate.	Cloudy	Cumulus clouds; hard rain on the Canadian, nights of 17th and 18th.
18	12 m	27.51	97	87	N.	...do...	...do...	Nimbus clouds; slight rain to the south.
18	6 p. m	27.5	88	80	S. E.	...do...	...do...	Cumulus clouds; rain on Staked Plains.
19	6 a. m	27.5	74	70	N.	...do...	...do...	Nimbus clouds; rained during the night.
19	9.30 p. m	27.4	69	66	N.	Light...	...do...	Rained at 1 p. m.
20	2.45 a. m	27.45	69	65	N.	Moderate.	...do...	Dew; stratus and nimbus clouds.
20	11.30 a. m.	27.53	78¾	77	S. E.	Light....	Clear....	
20	2 p. m	27.565	88	80	S. E.	...do...	...do...	
20	4.30 p. m	27.53	83½	77½	S. E.	Moderate.	...do...	
20	8.15 p. m	27.49	75	70	S. S. E.	Light....	Cloudy...	Dew.
20	10 p. m	27.49	74	69½	S. S. E.	...do...	Hazy.....	Dew.
21	3 a. m	27.45	67	66	S. S. E.	...do...	Clear....	
21	12 m	27.56	90¾	85	S.	Moderate.	...do...	Some clouds.
21	3 p. m	27.54	92¼	85½	S.	Strong...	...do...	Some clouds.
21	8 p. m	27.45	82	72	S.	Light....	...do...	Light stratus clouds in the horizon.
21	9 p. m	27.41	76¼	68	S.	...do...	...do...	
21	10.30 p. m.	27.4	73	66	S.	Very light.	...do...	No dew.
22	2.30 a. m	27.357	70	66¼	S.	...do...	...do...	Dense fog until 7 a. m.

APPENDIX A.—METEOROLOGICAL OBSERVATIONS.

Day	Time	Bar.			Wind dir.	Wind force	Sky	Remarks
22	12 m	27.675	97½	89	S.	Strong	...do...	Nimbus clouds floating.
22	4 p. m	27.60	96	88½	S.	...do...	...do...	Nimbus and stratus clouds light in the west.
22	7 p. m	27.62	85	80	S. W.	Mte.	Cloudy	
22	8.30 p. m	27.63	83	74	W.	Violent	...do...	Raining and blowing very violently. Violent storm during the night, with thunder and lightning.
23	3 a. m	27.61	77	69½	N. W.	Light	...do...	
23	8.30 a. m	27.725	73	67	N. E.	Mte.	...do...	Rained since 5 a. m.
23	10 a. m	27.73	73	68½	N. E.	Strong	...do...	Rain just ceased; nimbus clouds.
23	1.30 p. m	27.79	73	68	N. E.	Mte.	...do...	Nimbus clouds; raining slightly; has been raining at intervals.
23	7 p. m	27.825	70½	63½	N. E.	Strong	...do...	Nimbus clouds.
23	9.30 p. m	27.84	67	63	E. N. E.	...do...	...do...	Nimbus clouds.
24	3 a. m	27.89	66	65	E. N. E.	Moderate	...do...	Nimbus clouds.
24	1 p. m	28.02	76	74½	E. N. E.	Light	...do...	
24	5 p. m	28.00	76	68	E. N. E.	...do...	...do...	
24	7 p. m	28.00	73	64	E. N. E.	...do...	Clear	Stratus clouds in the southwest.
24	10 p. m	28.00	62	59	N. N. E.	...do...	...do...	Dew.
25	3 a. m	28.105	59	56	N. N. E.	...do...	...do...	Heavy dew.
25	2 p. m	28.075	73	74	S.	...do...	...do...	
25	5 p. m	28.06	79	71	S.	...do...	...do...	No dew.
25	7 p. m	28.05	72	66	S.	...do...	...do...	Dew.
25	9 p. m	27.97	67	62	E. N. E.	...do...	...do...	Floating clouds.
26	3 a. m	28.02	59	56	S. S. E.	...do...	...do...	Floating clouds in larger quantity.
26	calm	27.975	85	79½	S. S. E.	Mte.	...do...	Floating clouds.
26	2.30 p. m	27.9	92	87	S. W.	...do...	...do...	Nimbus clouds in southwest and west; thunder.
26	5.30 p. m	27.89	90	80		Light	Cloudy	
26	7 p. m		82	73				
26	9.30 p. m	27.88	78	69	W. S. W.	Mte.	...do...	Violent wind and rain; thunder and lightning about 8 p. m.
27	3 a. m	27.825	67½	62	W. S. W.	...do...	Clear	Rained at intervals until midnight.
27	12m	28.06	83	81	N.	Light	...do...	

METEOROLOGICAL OBSERVATIONS—Continued.

Date.	Hour.	Barometer.	Thermometer.		Wind.		Weather.	Remarks.
			Attached.	Detached.	Direction.	Intensity.		
1852.		"	°	°				
June 27	2 p. m......	28.08	90	83	N.	Moderate.	Clear......	Floating clouds.
27	6.30 p. m...	28.05	82	74	N.	...do......	...do......	Floating stratus clouds.
27	9.30 p. m...	28.045	72½	67	N. N. E.	Very light.	Cloudy.....	Nimbus clouds; lightning in the west and northwest.
28	3 a. m......	27.98	68	62	N. N. E.	...do......	Clear......	Rained during the night.
28	11.30 a. m .	27.68	82	83	N. N. E.	Moderate.	...do......	Clouds; lightning in the northwest and west.
28	10.30 p. m..	27.6	76	70	S. W.	...do......	...do......	
29	4 a. m......	27.5	75	73	S. E.	Strong.....	...do......	
29	12 m.......	27.95	104	100	W.	Moderate.	...do......	
29	6 p. m......	27.85	98	86	S. E.	...do......	...do......	
30	4.30 a. m...	27.85	82	80	S.	...do......	...do......	
30	12 m.......	27.95	108	103	S. S. E.	...do......	...do......	
30	6 p. m......	27.9	104	91	W.	...do......	Cloudy.....	
July 1	4 a. m......	27.85	74	69	W.	...do......	...do......	
1	12 m.......	28.00	112	104	W.	...do......	Clear......	
1	6 p. m......	27.925	104	91	S. S. E.	...do......	Cloudy.....	
2	4.30 a. m...	27.95	74	66	N. W.	...do......	...do......	
2	12 m.......	28.1	92	90	N. E.	...do......	Clear......	Cumulus clouds.
2	6 p. m......	28.0	88	80	S. E.	...do......	Cloudy.....	
3	4.30 a. m...	27.95	76	71	S. E.	Calm	...do......	Nimbus clouds.
3	12 m.......	27.95	91	89	S.	Strong.....	Clear......	

APPENDIX A.—METEOROLOGICAL OBSERVATIONS. 133

3	10 p. m.	27.9	79	74	s.	...do...	Cloudy	
4	3 a. m.	27.9	70	66	s.	Mte.	...do...	
4	12 m.	27.95	101	90	s.	Light	Clear	
4	9 p. m.	27.9	78	72	s.	...do...	...do...	
5	1.30 a.m.	27.875	78½	74	s.	...do...	...do...	No dew.
5	12 m.	28.185	100	92	s.	Strong	...do...	
5	3 p. m.	28.2	105	94	s.	...do...	...do...	
5	7.30 p. m.	28.125	86	80	s.	Light	...do...	
5	8.30 p. m.	28.11	82	80	s.	Strong	...do...	No dew.
6	1 a. m.	28.165	82	76	s.	Light	...do...	No dew.
6	10.30 a. m.	28.28	85	83	s.	Strong	...do...	
6	3 p. m.	28.28	102	90	s.	...do...	...do...	No dew; lightning in the north; many meteors.
6	7.30 p. m.	28.215	84	82½	s.	Mte.	...do...	No dew.
6	8.30 p. m.	28.225	83	82	s.	...do...	...do...	No dew; a few clouds floating.
7	1.30 a. m.	28.225	80	74	s.	Light	...do...	Cloudy from 5 a. m. to 10 a. m.; nimbus clouds.
7	3 a. m.	28.22	75	73	s.	...do...	...do...	
7	8 a.m.	28.27	100	90	s.	Strong	...do...	No dew.
7	2 p. m.	28.26	102	92	s.	Violent	...do...	Nimbus clouds; thunder and lightning in N; a few drops of rain.
7	9 p. m.	28.205	85½	82	s.	...do...	...do...	
8	12.30 a. m.	28.2	81	77	s.	Strong	...do...	Nimbus clouds, lightning in the S. W.; luminous appearance in N. (turn light.)
8	12 m.	28.52	101	93	s.	...do...	...do...	
8	4.30 p. m.	28.475	105	90	s.	...do...	...do...	
8	7 p. m.	28.43	88	83	s.	Light	Cloudy	
8	9 p. m.	28.525	87	79	s.	Very light	...do...	Nimbus and cirrus clouds; strong wind during the night.
9	1.30 a. m.	28.43	82	77	s.	Moderate	...do...	
9	10 a. m.	28.725	91	87	s.	...do...	Clear	
9	12 m.	28.73	104½	91	s.	Strong	...do...	

METEOROLOGICAL OBSERVATIONS—Continued.

Date.	Hour.	Barometer.	Thermometer. Attached.	Thermometer. Detached.	Wind. Direction.	Wind. Intensity.	Weather.	Remarks.
1852.		"	°	°				
July 9	4 p. m.	28.645	104½	86	S.	Strong	Clear	
9	8 p. m.	28.56	84	74	S.	Very light	...do...	No dew.
9	9 p. m.	28.55	81	74	S.	...do...	...do...	Do.
9	10.30 p. m.	28.55	77	70	S.	Light	...do...	Do.
10	2.30 a. m.	28.53	68	62	S.	Very light	...do...	
10	12 m.	28.775	103	94	S.	Strong	...do...	
10	4.45 p. m.	28.715	109	98	S.	...do...	...do...	Light clouds in the horizon N. by N. W.
10	7.30 p. m.	28.7	92	86	S.	Light	...do...	No dew.
10	9 p. m.	28.74	88	82	S.	Moderate	...do...	
11	2.30 a. m.	28.725	80	73	S.	...do...	...do...	
11	11 a. m.	28.96	100½	96	S.	Light	...do...	
11	12 m.	29.15	110¼	99	S.	Very light	...do...	
11	4.30 p. m.	28.9	106	92	S.	Moderate	Cloudy	
11	7.30 p. m.	28.875	89	82	S.	Very light	Clear	
11	9.30 p. m.	28.885	85	79½	S.	Moderate	...do...	No dew.
12	1 a. m.	28.86	80	74	S.	Light	...do...	
12	11.30 a. m.	29.00	108½	95	S.	...do...	...do...	
12	3 p. m.	28.965	110	95	S.	...do...	...do...	Cumulus clouds.
12	7.30 p. m.	28.88	90	83½	S.	...do...	Cloudy	No dew; nimbus clouds; lightning in the north and west.
12	9.30 p. m.	28.905	87	81½	S.	Moderate	...do...	

APPENDIX A.—METEOROLOGICAL OBSERVATIONS.

Day	Time	Pressure			Temp	Wind Dir	Wind	Sky	Remarks
13	4.30 a. m.	28.925	80		76	S.	do	do	Nimbus clouds; raining slightly.
13	6 a. m.	28.922	77		72½	S.	Light	do	Nimbus clouds; rained for about one hour.
13	10 a. m.	29.25	85		78	S.	Strong	do	Stratus clouds.
13	12 m.	29.2	95		84	E. N. E.	Light	do	Do. do.
13	4 p. m.	28.975	95		83	N. E.	Mod.	Clear	
13	7.30 p. m.	28.93	83		74	E. N. E.	Light	do	Stratus beds in the west.
13	10.30 p. m.	28.925	74		68	E. N. E.	Calm	Clear	No dew.
14	3 a. m.	28.92	73		68	E. N. E.	do	One half clear	Slight dew
14	11 a. m.	29.00	98½		92	N. E.	Light	¼ clear	Cirrus clouds.
14	4 p. m.	28.92	95		86	N. E.	Mod.	½ clear	Cirrus clouds; from 12½ to 1½ p. m. rained moderately.
14	7.30 p. m.	28.875	82		76	N. E.	Calm	do	Cumulus clouds.
14	9.30 p. m.	28.875	79½		74	N. E.	Very light	⅘ clear	No dew; cirrus; meteors.
15	1.30 a. m.	28.85	76		69½	S.	do	Clear	Slight dew; almost perfectly calm.
15	10 a. m.	28.925	101½		90	S.	do	⅛ clear	Cirrus clouds.
15	11 p. m.	28.925	103		100	S.	do	do	Do. do.
15	3 p. m.	28.87	105		94	S.	Moderate	½ clear	Nimbus and cirrus clouds; thunder in the southeast.
15	6 p. m.	28.8	84		77	N. E.	do	⅛ clear	1 Cirrus clouds.
15	9 p. m.	28.825	78		73½	N. E.	Light	½ clear	1 Cirrus clouds; no dew; luminous appearance in northwest.
16	2 a. m.	28.8	74		69	E.	do	Clear	Slight dew.
16	11 a. m.	28.85	87		82	S.	do	⅞ clear	
16	2½ p. m.	28.8	92		85	S.	Moderate	½ clear	
16	7.30 p. m.	28.725	80		76	E.	Light	Clear	
16	9 p. m.	28.73	77		72	S.	do	do	No dew.
17	3 a. m.	28.72	74		69½	S.	do	do	Heavy dew.
17	11 a. m.	28.975	88		86	S. E.	do	⅛ clear	Cumulus clouds.
17	3 p. m.	28.955	99		84	N. E.	Moderate	⅞ clear	Do. do.
17	5 p. m.	28.92	93		84	N. E.	do	Clear	
17	7.30 p. m.	28.875	84		74	N. E.	Light	do	
17	10 p. m.	28.875	75		68¼	E.	Very light	do	Slight dew.

METEOROLOGICAL OBSERVATIONS—*Continued.*

Date.	Hour.	Barometer.	Thermometer.		Wind.		Weather.	Remarks.
			Attached.	Detached.	Direction.	Intensity.		
1852.		"	°	°				
July 18	2. 30 a. m.	28. 87	72½	67	E.	Very light.	Clear.	Very heavy dew.
18	12 m.	28. 95	98	89	S. S. E.	Moderate.	$\frac{9}{10}$ clear.	
18	6 p. m.	28. 85	84½	79	S. S. E.	Light.	Clear.	
18	7. 30 p. m.	28. 825	80	73	S. S. E.	Calm	$\frac{9}{10}$ clear.	
19	3 a. m.	28. 86	69	66	S. S. E.	...do.	Clear.	Heavy dew.
19	12 m.	29. 08	86	81	S.	Moderate.	½ clear.	
19	3 p. m.	29. 07	90½	85	S.	....do.	⅓ clear.	
19	5. 30 p. m.	29. 02	85	79	S.	....do.	¾ clear.	
19	7. 30 p. m.	29. 00	80	73	S.	Light.	....do.	
19	10 p. m.	29. 00	74½	70	S.	Very light.	Clear.	Slight dew.
20	2 a. m.	28. 975	71	66	S.	Light.	...do.	Heavy dew.
20	12 m.	29. 075	85	82½	S. S. W.	Moderate.	¼ clear.	Stratus and cumulus clouds.
20	4 p. m.	29. 06	91	82	S.	....do.	¾ clear.	Cumulus clouds.
20	7. 30 p. m.	29. 025	78	71	S.	Light.	¼ clear.	
20	9. 30 p. m.	29. 025	72½	68	S.	...do.	Clear.	No dew.
21	2. 30 a. m.	29. 00	65	60	S.	Calm	½ clear.	Moderately heavy dew; stratus and cumulus clouds in the N.; nimbus clouds; thunder and lightning S.
21	12 m.	29. 125	87	82	S.	Moderate.	¾ clear.	Nimbus clouds; rained [illegible] hard from 5½ a. m. to 7 a. m.; came up from the south.
21	3 p. m.	29. 12	90	84	S.	....do.	....do.	Stratus and cumulus clouds.

21	9.30 p. m.	29.07	72	68	S.	Calm	$\frac{9}{10}$ clear	Cds; no dew.
22	1.30 a. m.	29.025	67	62	S.	Very light.	½ clear	Stus clouds; heavy dew.
22	1.30 p. m.	29.075	78	75	S.	Moderate.	Cloudy	Nus clouds; rained from 12½ p. m. to 1 p. m.
22	4.30 p. m.	29.025	77½	72	S.	Very light.	...do	Nimbus clouds.
22	10 p. m.	29.02	72	68	S	Calm	½ clear	Do. do.
23	5.30 a. m.	29.03	72	68	S.	Moderate.	Cloudy	Nimbus clouds; rned hard during the night ard early in the morning.
23	7.30 a. m.	29.07	74½	71	S.	....do	...do	Nus cls.
23	10 a. m.	29.13	91½	88	S.	....do	¼ lear	Nius clouds.
23	12 m.	29.075	96	94	S.	Strong	⅓ clear	Cumulus clouds.
23	7.30 p. m.	28.975	78	72	S.	Calm	$\frac{9}{10}$ clear	Do. do.
23	0.30 p. m.	28.95	73	68	S.	Light.	do.	Stratus luds.
24	3 a. m.	28.93	72	68	S.	...do	½ lear	Nius and stratus clouds; dew.
24	1.30 p. m.	29.01	101	88	S.	...do	¾ lear	
24	6.30 p. m.	28.93	89	83	S.	...do	do.	Stratus clouds.
24	9.30 p. m.	28.92	81	76	S.	...do	½ clear	Stratus clouds; no dew.
25	2.30 a. m.	28.86	77	72	S.	Calm	Cloudy	Nimbus clouds; rained from 8½ to 9 a. m., from 11½ to 1 p. m., and from 2½ to 3 p. m.
25	2 p. m.	28.95	80	75	S.	...do	...do	
25	10 p. m.	28.91	74	68	S. E.	Light.	¾ clear	
26	4 a. m.	28.91	71	62	S. E.	Calm	Cloudy	Nimbus and rstus clouds.
26	2 p. m.	29.225	88	85	E.	Light.	½ clear	Kds; rned from 8½ to 9 a. m., from 10½ to 11½ a. m., and from 1 to 1½ p. m.
26	4 p. m.	29.210	87	82	E.	...do	....do	Cus luds; rd from 3 to 3½ p. m.
26	9.30 p. m.	29.200	76	72	E.	Calm	....do	Slight dew; dense fog in the bottom.
27	3 a. m.	29.200	72	65	E.	...do	....do	Heavy dew.
27	1.30 p. m.	29.240	95	88	S.	Light.	¾ clear	Cumulus luds; slight rain at 12 m.
27	4.30 p. m.	29.230	97	84	S.	...do	½ clear	C ulus and stus luds.

APPENDIX B.

COURSES AND DISTANCES.

TABLE OF COURSES AND DISTANCES ON THE ROUTE OF CAPT. MARCY'S EXPLORATION OF RED RIVER.

APPENDIX B.—COURSES AND DISTANCES.

Table of courses and distances on the route of Captain Marcy's expedition to the sources of Red river.

Date.	No. of camp. Arrival.	Course.	Bearing.	No. of revolutions of odometer.	Distance in statute miles.	Remarks.
1852. May 7	......	S. E. 6° S.	39	......	......	Total distance travelled, 6,247 revolutions = 16 miles 5 yards.
		N. 7 W.	187			
		N. 54 W.	234			
		N. 66 W.	246			
		N. 88 W.	268			
		N. 60 W.	240			
		N. 30 W.	210			
		N. 85 W.	265			
		S. 80 W.	280			
		N. 28 W.				
8	......	N. 78 W.	208	......	......	Total distance travelled, 2,292 revolutions = 5 miles 1,524 yards.
		N. 10 W.	258			
		N. 78 E.	190			
		N. 52 E.	112			
		N. 34 E.	142			
11	......	N. E. 11 N.	146	......	......	Total distance travelled, 1,567 revolutions = 4 miles 11 yards.
		N. W. 15 W.	240			
12	......	N. E. 2 N.	223	1,090	mls. 2 799	Total distance travelled, 4,664 revolutions = 11 miles 92.
		W. 6 S.	276	483	1 241	
		S. W. 8 W.	307	786	2 019	
		S. W. 1 W.	314	208	0 534	

APPENDIX B.—COURSES AND DISTANCES.

13	...1...	S. W. 15 W.	300	2,097	5	386	Total distance travelled, 434 revolutions = 1 mile 115.
16	...2...	S. W. 5 S.	320	434	1	115	Total distance travelled, 5,625 revolutions = 14 miles 228.
		W. 6 S.	276	548	1	407	
		W. 27 S.	297	1,442	3	704	
		W. 10 S.	280	994	2	553	
		W. 20 S.	290	1,534	3	940	
		W.	270	1,107	2	843	
17	...3...	N. W. 5 W.	230	903	2	319	Total distance travelled, 4,341 revolutions = 11 miles 172.
		S. W. 17 W.	298	1,462	3	755	
		S. W. 15 W.	300	526	1	351	
		W. 10 S.	280	527	1	353	
		N. W. 15 W.	240	923	2	370	
18	...4...	N. W. 15 W.	240	869	2	233	Total distance travelled, 5,684 revolutions = 14 miles 34.
		W. 20 N.	250	980	2	517	
		W. 15 N.	255	618	1	597	
		W. 10 N.	260	611	1	597	
		N. W.	225	856	2	198	
		W. 10 N.	260	539	1	384	
		W. 12 N.	258	463	1	189	
		N. W. 5 W.	230	748	1	921	
19							Halted in Camp No. 4, on Sink creek.
20	...5...	W. 19 N.	251		3	00	Total distance travelled, 4 miles 405.
		N. 10 W.	190		1	405	
21	...6...	N. W. 15 N.	210		0	250	Total distance travelled, 3 miles 41.
		S. W. 8 S.	323		0	500	
		N. W. 5 W.	230		1	500	

COURSES AND DISTANCES—*Continued.*

Date.	No. of camp. Arrival.	Course.	Bearing.	No. of revolutions of odometer.	Distance in statute miles.	Remarks.
1852. May 21	6	N. E. 20 N.	155°		*mls.* 1 160	
22	7	W. 10 N.			2 884	Total distance travelled, 24 miles 163.
		N. W. 10 W.			7 798	
		N. W. 5 N.			3 135	
		N. 20 W.			1 142	
		N. N. W.			9 209	
26	8	E.	90		0 852	Total distance travelled, 852.
27	9	E. 22° N.	112		1 00	Total distance travelled, 1 mile 295.
		N. E. 17 E.	118		0 295	
29	10	W. 6 S.	276	424	1 084	Total distance travelled, 6 miles 09153.
		N. 5 E.	175	233	0 598	
		N. W. 17 N.	208	521	1 338	
		N.	180	1,195	3 069	
30	11	N. 10 W.	190	1,031	2 648	Total distance travelled, 3,198 revolutions = 8 miles 225359.
		N. 20 E.	160	384	0 986	
		N. 20 W.	200	526	1 351	
		N. 10 W.	190	322	0 827	
		N. E. 5 E.	130	628	1 623	
		N. E. 5 N.	140	307	0 788	

APPENDIX B.—COURSES AND DISTANCES. 143

31	..12..	N. W. 5 N.	220	1,638	4 207	Total distance travelled, 3,430 revolutions = 8 miles 820395.
		N. W. 15 N.	210	280	0 719	
		N. W.	225	625	1 615	
		N.	180	803	2 062	
		N. 5 E.	175	84	0 215	
June 1	..13..	N. 20 W.	200	1,296	3 585	Total distance travelled, 2,470 revolutions = 6 miles 6.
		N. W. 5 W.	230	235	0 603	
		W. 8 N.	262	939	2 412	
2	..14..	N. W. 7 W.	232	1,160	2 980	Total distance travelled, 4,346 revolutions = 11 miles 181.
		N. W. 5 N.	220	717	1 841	
		N. W. 9 W.	234	697	1 790	
		N. W. 11 N.	214	620	1 602	
		N. W. 5 N.	220	498	1 279	
		N. W. 17 N.	208	654	1 689	
3	..15..	N. W. 5 N	220	1,009	2 591	Total distance travelled, 2,945 revolutions = 7 miles 574.
		N. E. 7 N.	142	969	2 490	
		W. 5 N.	265	626	1 618	
		W.	270	341	0 875	
4	..16..	W.	270	1,061	2 726	Total distance travelled, 5,468 revolutions = 14 miles 054.
		W. 8 S.	278	419	1 076	
		S. W. 9 S.	324	494	1 269	
		W.	270	1,063	2 731	
		N. 30 W.	210	986	2 532	
		N.	180	638	1 648	
		N. 30 W.	210	807	2 072	
5	..17..	N. 15 W.	195	616	1 592	Total distance travelled, 3,442 revolutions = 8 miles 830.
		N. W. 5 N.	220	619	1 600	
		N. W.	225	1,109	2 848	

144   APPENDIX B.—COURSES AND DISTANCES.

COURSES AND DISTANCES—*Continued.*

Date.	No. of camp. Arrival.	Course.	Bearing.	No. of revolutions of odometer.	Distance in statute miles.	Remarks.
1852. June 5	17	N. W. 17 N. N. 15 W.	208 195	493 605	*mls.* 1 266 1 564	
6	18	W. 20 N. W. 10 N. N. W. 15 W. W. 15 N.	250 260 240 255	1,048 671 960 819	2 692 1 733 2 467 2 104	Total distance travelled, 3,498 revolutions = 8 miles 996.
7	19	N.	180	459	1 179	Total distance travelled, 459 revolutions = 1 mile 179.
8	20	N. W. 5 N. S. W. 15 W. N. W. 5 N. N. W. 20 N. N. W. 18 W.	220 300 220 205 238	465 688 972 802 246	1 195 1 777 2 496 2 060 0 632	Total distance travelled, 3,173 revolutions = 8 miles 16.
9	21	W. 14 S. S. W. 1° S. S. W. 5 S. N. W. 20 N. N. W. 2 N.	284 316 320 205 223	1,156 781 343 590 649	2 969 2 046 0 881 1 525 1 677	Total distance travelled, 3,519 revolutions = 9 miles 058.
10	22	W. 6 S. W. 18 N.	276 252	1,162 574	2 985 1 474	Total distance travelled, 4,127 revolutions = 10 miles 601.

APPENDIX B.—COURSES AND DISTANCES. 145

11	...23...	N.W. 13 N.	212		764	1 963	
		W. 10 N.	260		1,287	3 306	Total distance travelled, 4,243 revolutions = 10 miles 894.
		N.W. 13 N.	212		340	0 873	
12	...24...	W.	270		1,350	3 467	
		W. 5 N.	265		1,687	4 333	Total distance travelled, 3,480 revolutions = 8 miles 938.
		W. 16 S.	286		981	2 519	
		N.W. 15 W.	240		224	0 575	
14	...25...	W. 5 N.	265		1,476	3 791	
		W. 8 S.	278		431	1 107	Total distance travelled, 4,390 revolutions = 11 miles 275.
		W. 5 N.	265		353	0 906	
		N.W. 15 W.	240		1,220	3 134	
		S.W.	315		898	2 306	
		W. 10 S.	280		1,208	3 103	
		W. 3 S.	273		552	1 418	
		S.W.	300		1,142	2 933	
		S.W. 17 W.	288		590	1 515	
15	...26...	W.	270		822	2 101	
		S. 10 W.	350		891	2 288	Total distance travelled, 4,044 revolutions = 10 miles 387.
		W. 10 N.	260		799	2 052	
		S.W. 15 W.	300		1,050	2 697	
					486	1 249	
16	...27...	W. 10 S.	280		1,017	2 612	
		W.	270		305	0 783	
		S.W. 15 S.	330		1,302	3 344	Total distance travelled, 5,851 revolutions = 15 miles 037.
		W. 6 S.	276		674	1 741	
			320		401	1 030	
			350		976	2 507	
					,176	3 020	

10

APPENDIX B.—COURSES AND DISTANCES.

COURSES AND DISTANCES—*Continued.*

Date.	No. of camp. Arrival.	Course.	Bearing.	No. of revolutions of odometer.	Distance in statute miles.	Remarks.
1852. June 17	......28......	S. E. 10 E.	55°	1,132	mls. 2 706	Total distance travelled, 1,132 revolutions = 2 miles 706.
20	......29......	S. S. 10 E. S. 10 W.	360 10 350	2,808 983 608	7 213 2 525 1 572	Total distance travelled, 4,399 revolutions = 11 miles 310.
21	......30......	S. 20 W. S. E. 10 S. S. 10 E. S. 10 E. S. E. 15 S. S. W. 5 W.	340 35 360 10 360 30 310	945 851 415 556 641 838 311	2 427 2 262 1 065 1 429 1 756 2 152 0 788	Total distance travelled, 4,587 revolutions = 11 miles 888.
22	......31......	S. S. 20 W. S. E. E.	360 340 45 90	2,358 837 999 695	6 036 2 150 2 560 1 785	Total distance travelled, 4,889 revolutions = 12 miles 537.
23	......32......	E.	90	1,632	4 192	Total distance travelled, 1,632 revolutions = 4 miles 192.
24	......33......	S. E. 10° S. S. 10 E. S. E. 20 S.	35 10 25	604 554 1,681	1 561 1 423 4 318	Total distance travelled, 5,179 revolutions = 13 miles 303.

APPENDIX B.—COURSES AND DISTANCES. 147

25	....34....	S.	360	1,174	3 025	
		S. E.	45	705	1 811	
		S.	360	461	1 185	Total distance travelled, 5,807 revolutions = 14 miles 916.
26	....35....	S. 10 W.	340	1,217	3 015	
		N. W. 15 N.	210	579	1 487	
		W. 10 S.	280	975	2 502	
		S. 10 W.	350	260	0 667	
		W.	270	843	2 164	
		S. W.	315	1,933	4 964	Total distance travelled, 3,578 revolutions = 9 miles 2.
27	....36....	W. 20 N.	250	789	2 026	
		S.	360	842	2 162	
		W. 10 S.	280	1,043	2 676	
		W. 10 N.	260	904	2 323	Total distance travelled, 4,557 revolutions = 11 miles 705.
28	....37....	S. W. 10 W.	305	936	2 403	
		S. 5 W.	355	876	2 249	
		S. 10 E.	10	323	0 829	
		S. W. 10 S.	325	1,444	3 709	
		S. E.	45	159	0 408	
		S.	360	819	2 103	Total distance travelled, 3,496 revolutions = 8 miles 98.
29	....38....	S. 6 W.	354	1,247	3 203	
		S. 10 W.	350	423	1 085	
		S. W.	315	993	2 55	
		W.	270	519	1 333	
		S. 20 W.	340	314	0 802	
		N. E. 20 N.	155	763	1 960	
		N. E. 20 E.	115	778	1 998	
		N. E.	130	512	1 314	
		N. E. 5 E.	160	640	1 653	
		N. 20 E.				Total distance travelled, 2,693 revolutions = 6 miles 917.

APPENDIX B.—COURSES AND DISTANCES.

COURSES AND DISTANCES—*Continued.*

Date.	No. of camp. Arrival.	Course.	Bearing.	No. of revolutions of odometer.	Distance in statute miles.	Remarks.
1852. July 2	39	E.	90°	883	*mls.* 2 267	Total distance travelled, 883 revolutions = 2 miles 267.
4	40	N. E. 15 E.	120	753	1 935	Total distance travelled, 5,099 revolutions = 13 miles 097.
		E.	90	475	1 218	
		S. E.	45	626	1 617	
		N. 5 E.	175	270	0 693	
		E. 20 N.	110	460	1 180	
		N. E. 20 E.	115	669	1 729	
		E.	90	1,846	4 741	
5	41	E. 10 N.	100	829	2 129	Total distance travelled, 6,055 revolutions = 15 miles 653.
		E. 10 S.	80	1,067	2 741	
		E. 20 S.	70	1,500	3 853	
		E.	90	609	1 574	
		N. 10 E.	170	1,132	2 906	
		S.	360	918	2 356	
6	42	E. 10 N.	100	1,028	2 639	Total distance travelled, 4,995 revolutions = 12 miles 83.
		N. E. 15 E.	120	656	1 694	
		N. E. 10 E.	125	1,083	2 780	
		E. 10 N.	100	1,115	2 861	
		E. 10 S.	80	335	0 860	
		N. E.	135	778	1 997	

APPENDIX B.—COURSES AND DISTANCES.

		Course				Total
7	43	N. E. 5 E.	130	2,295	5 894	Total distance travelled, 6,208 revolutions = 16 miles 046.
		E. 10 S.	80	691	1 775	
		E.	90	538	1 381	
		N. E. 15 E.	120	973	2 497	
		E. 10 N.	100	1,049	2 693	
		E. 20° S.	70	662	1 711	
8	44	E. 10 S.	80	535	1 373	Total distance travelled, 6,348 revolutions = 16 miles 418.
		E. 20 S.	70	2,411	6 191	
		E.	90	1,320	3 389	
		S. E. 15 E.	60	1,373	3 524	
		S. E. 5 E.	50	709	1 821	
9	45	N. E.	135	600	1 551	Total distance travelled, 4,159 revolutions = 10 miles 683.
		E. 20 N.	110	742	1 905	
		E. 10 N.	100	465	1 192	
		N. E. 5 E.	130	994	2 553	
		N. E. 15 N.	50	795	2 042	
		E. 20 S.	70	562	1 446	
10	46	S. E.	45	1,962	5 037	Total distance travelled, 5,357 revolutions = 13 miles 76.
		S. E. 15 S.	30	1,205	3 093	
		S. E. 7 E.	52	836	2 146	
		S. E.	45	695	1 786	
		S. 10 W.	350	659	1 702	
11	47	E.	90	1,043	2 677	Total distance travelled, 4,316 revolutions = 11 miles 086.
		E. 20 N.	110	1,034	2 646	
		E. S. 15 E.	60	499	1 282	
		E.	90	1,740	4 469	
12	48	E. 5 N.	95	2,196	5 640	Total distance travelled, 5,604 revolutions = 14 miles 4.
		E. 20 N.	110	1,209	3 103	

COURSES AND DISTANCES—*Continued.*

Date.	No. of camp. Arrival.	Course.	Bearing.	No. of revolutions of odometer.	Distance in statute miles.	Remarks.
1852. July 12	48	N. E. 21 E. E. 20 N. S. E. 10 E.	° 114 110 55	735 692 772	*mls.* 1 888 1 779 1 982	
14	49	S. E. 15 E. E. 20 S. N. E. 5 E. N. 20 E. N. E. 5 E.	60 70 130 160 130	909 810 794 571 1,231	2 334 2 079 2 039 1 466 3 159	Total distance travelled, 4,315 revolutions = 11 miles 084.
15	50	E. 10 N. N. E. N. E. 5 N. N. E. E. 20 N.	100 135 140 135 110	1,916 784 663 473 234	4 919 2 014 1 712 1 212 0 600	Total distance travelled, 4,070 revolutions = 10 miles 454.
16	51	N. E. 15 N. N. E. 5 E. E. E. 10 N. E.	150 130 90 100 90	1,720 98 522 336 239	4 418 0 251 1 34 0 862 0 613	Total distance travelled, 2,915 revolutions = 7 miles 487.
17	52	E. E. 10 N.	90 100	1,844 451	4 734 1 157	Total distance travelled, 4,669 revolutions = 11 miles 993.

APPENDIX B.—COURSES AND DISTANCES.

		Course					Total
18	..53..	S. E. 15 E.	60	656		1	694	Total distance travelled, 3,231 revolutions = 8 miles 299.
		E. 5 N.	95	749		1	924	
		E. 20 N.	110	969		2	489	
19	..54..	N.	180	895		2	299	Total distance travelled, 2,482 revolutions = 6 miles 376.
		N. 20 E.	160	620		1	602	
		E. 20 N.	110	958		2	461	
		N. E.	135	758		1	947	
20	..55..	S. E. 5 S.	40	304		0	78	Total distance travelled, 3,602 revolutions = 9 miles 25.
		S. E.	45	801		2	056	
		S. E. 15 E.	60	1,377		3	536	
21	..56..	S. E. 5° S.	40	1,023		2	626	Total distance travelled, 3,855 revolutions = 9 miles 902.
		N. 5 E.	175	614		1	587	
		E. 10 N.	100	805		2	066	
		E. 15 N.	105	1,160		2	979	
22	..57..	N. E. 5 E.	130	943		2	42	Total distance travelled, 7,074 revolutions = 18 miles 17.
		E. 5 S.	85	2,194		5	634	
		E. 20 S.	70	718		1	843	
		S. E. 15 E.	60	2,099		5	92	
		N. E. 5 E.	130	617		1	93	
		E. 10 N.	90	395		0	05	
		E.	100	916		2	31	
			90	3,047		7	85	
24	..58..	S.	360	1,853		4	59	Total distance travelled, 4,957 revolutions = 12 miles 734.
		S. E.	45	564		1	49	
		S. E. 15 E.	60	945		2	26	
		E.	360	1,595		4	06	

COURSES AND DISTANCES—Continued.

Date.	No. of camp. Arrival.	Course.	Bearing.	No. of revolutions of odometer.	Distance in statute miles.	Remarks.
1852. July 25	....59....	S.	° 360		mls.	
		S. 10° E.	10	1,209	3 105	Total distance travelled, 5,905 revolutions = 15 miles 158.
		S. E. 15 E.	60	1,953	5 008	
		S. E. 5 S.	40	675	1 733	
		S. 10 E.	10	855	2 197	
				1,213	3 115	
26	....60....	N. E. 10 E.	125	543	1 393	Total distance travelled, 3,827 revolutions = 9 miles 887.
		N. E. 15 E.	120	1,784	4 583	
		N. E. 5 E.	130	969	2 488	
		N. E. 15 E.	120	531	1 423	
27	....61....	N. E.	135	737	1 892	Total distance travelled, 5,367 revolutions = 13 miles 803.
		S. E.	45	641	1 656	
		E. 10 S.	80	627	1 619	
		E. 15 S.	75	1,820	4 675	
		N. E. 10 E.	125	1,542	3 961	
28	....62....	N. E. 5 E.	130	3,736	9 596	Total distance travelled, 3,736 revolutions = 9 miles 596.

APPENDIX C.

MINERALOGY.

REPORT ON THE MINERALS COLLECTED: BY PROF. CHARLES UPHAM SHEPARD.

APPENDIX C.

REPORT ON THE MINERALS COLLECTED: BY PROF. CHARLES UPHAM SHEPARD.

AMHERST COLLEGE, *June* 1, 1853.

MY DEAR SIR: The following report relates to the specimens collected by Captain Marcy, and which, agreeably to your request, were submitted by me to a chemical and mineralogical examination.

Very respectfully and truly yours,
CHARLES UPHAM SHEPARD.

To President HITCHCOCK.

1. COPPER ORES—MARCYLITE.

The most interesting of these was a specimen of rather more than one ounce in weight, from the main or south fork of Red river, near the Witchita mountains. It is a black compact ore, strongly resembling the black oxide of copper from the Lake Superior mines, for which substance I at first mistook it. It was partially coated by a thin layer of the rare and beautiful atacamite, (muriate of copper of Phillips.) This is the first instance in which this species has been detected in North America. On subjecting the black ore to a close investigation, it proves to be a substance hitherto undescribed, and it affords me much pleasure to name it, in honor of the very enterprising and successful explorer to whom mineralogy is indebted for the discovery, *Marcylite*. It is massive and compact; fracture even; color black; opaque; lustre none; hardness equals that of calcite, or 3 of the mineralogical scale; sectile streak shining; powder light grayish black; specific gravity, 4.0 to 4.1. In small fragments it melts in the heat of a candle, to the flame of which it imparts a rich blue and green color. This is especially striking when a blow-pipe is employed. The slightest heat of the instrument suffices for the fusion of the ore. The chloride of copper is volatilized, and spreads over the charcoal support, from which the splendid green color rises also. On directing the flame of the candle against it, the mass, or assay, remains for some time fluid, continuing to give the color as at first, till finally the green and blue tinge declines, and at last disappears altogether; after which the globule swells out into large bubbles and suddenly collapses, and this repeatedly for a number of times, (ten or fifteen,) when it seems to be pure copper. In cooling, however, a thin, light steel-gray pellicle forms upon its surface, which separates by a slight blow with the hammer, revealing a globule of pure copper within. This coating, on being fused with borax, gives rise to a colorless glass, with brilliant points of metallic copper adhering to the support of subjacent charcoal: a fragment heated in a small glass tube before the blow-pipe, enters into fusion and evolves much moisture, which contains traces of hydrochloric acid. The powdered mineral is almost wholly dissolved by ammonia, and the black powder which remains is slowly taken up by warm nitric acid, with the separation of traces of silica. Sulphuric acid dissolves the mineral, with the extrication of hydrochloric acid. Analysis gave the following as the composition of the ore:

Copper	54.30
Oxygen and chlorine	36.20
Water	9.50
	100.00

With traces of silica.

The above is undoubtedly a very valuable ore for copper, as it is very rich in metal, and easy of reduction in the furnace. Numerous specimens of the same ore, but very impure from an admixture of fine sand, were embraced in the collection, as coming from Copper creek, four miles from Cache creek. They were in the form of flattened, irregular discs, about two inches across and half an inch thick, having their surfaces coated by malachite (carbonate of copper) in a pulverulent condition. Along with the above, also, were found similarly shaped masses of an impure black oxide of copper, (coated by malachite,) which had the following composition:

Copper (with traces of iron)	35.30 to 40.00
Silica	30.60
Oxygen and water	34.10
	100.00

It is fusible before the blow-pipe, but does not tinge the flame blue or green. The fused mass bubbles up for a time, and finally yields a globule of copper with a thick crust, which is black, and feebly attracted by the magnet. A copper ore of the average characters of these flattened masses would yield from 33 to 35 per cent. in the large way.

Still another variety of copper ore is ticketed "June 3, Gypsum Bluff." It consists of numerous small fragments of a friable, fine-grained white sandstone, much mixed up with a pulverulent malachite, and occasionally presenting specks of black oxide of copper. Taken as a whole, I should judge that it might be a 5 per cent. ore.

Another variety still of copper ore, some stones, labelled "May 16 first day from Cache creek." They consist of a calcareous amygdaloid, through which are interspersed black oxide of copper and stains of malachite. Its value for metal would not exceed that of the variety last mentioned.

"May 17" refers to a compact grayish white limestone, much mottled with red. It contains druses of calcite, fibres of mesotype, and stains of black oxide of copper. It belongs, like the last, to the trappean family of minerals.

2. MANGANESE ORE.

"Copper creek, third day from Cache creek." An impure ore. When treated with warm hydrochloric acid, it evolves chlorine gas. It contains much silica, and some peroxide of iron, with 15.75 per cent. of water and about 10 per cent. binoxide of manganese. The specimen is imperfectly foliated, and, in places, is columnar. It is porous, and of a black color, resembling black oxide of copper. It was tested both for copper and cobalt, without detecting either.

3. IRON AND TITANIAN SANDS.

"July 18, Cache creek, foot of cliff." This is a heavy, rather coarse black sand; more than half of which consists of magnetic iron, the remainder being titaniferous iron. It is remarkable for its purity in these two minerals, the most careful search not resulting in the discovery of other minerals mingled with it, if we except a few grains of quartz labradorite and epidote. An ineffectual examination of it was had for tin and gold. Another specimen, collected July 16, was tested with a similar result.

4. OTHER MINERALS.

"May 31." Labradorite in numerous specimens. Its color is a dark pearl-blue, or gray; it does not fire the iridescent reflexions. From the size and purity of the masses, it would appear to be a very abundant mineral, even if it does not amount to a rock, throughout the region of the Witchita mountains.

Specimens were collected, bearing the same date, of a red cellular limestone, which may have originated, if we suppose a soft ferruginous clay to have been parted off by meshes, or cell-walls, of calcareous matter, and the clay to have been subsequently washed away, or in some manner mostly removed.

"July 15, base of Witchita mountains." Reddish septaria, or a mixture of peroxide of iron and calcite, traversed by veins of pure calcite, surfaces of the masses somewhat botryoidal. A singular variety of cellular quartz, said to have occurred in veins in the Witchita mountains, was carefully examined for gold, but without the detection of a trace of the precious metal, notwithstanding some of the specimens, from the presence of hydrated peroxide of iron and iron pyrites, looked very promising for gold.

5. SOILS.

"Sub-soil Cache creek, May 14; the same as that found about the Witchita mountains." The sample had been kneaded by the hand into a ball. Its color was reddish brown; it contained no organic matter. Analysis gave the following result:

Silica, (including some fine feldspathic grains)	82.25
Peroxide of iron	2.65
Alumina	0.55
Carbonate of lime	5.40
Carbonate of magnesia	1.70
Water (hygrometric moisture)	5.50
Sulphate of lime and carbonate of potash	traces
	98.05

This soil contains no perceptible traces of chlorine, or any other sulphate besides that of lime. It would appear to have an excellent constitution, as a sub-soil, for the cultivation of the grain crops, as well as for cotton. It is eminently a calcareous soil, and probably has a sufficiency of potash present also; but the quantity of the material did not enable me to determine the proportion of this constituent.

"Sub-soil, June 3." This sub-soil is fine grained, and has a clayey appearance. Its color is a deep red. Little fragments of gypsum may be detected scattered through its mass. It has the following rather unusual composition:

Silica	79.30
Peroxide of iron	8.95
Alumina	1.50
Carbonate of lime	1.10
Sulphate of lime, with strong traces of sulphate of soda and chloride of sodium	4.65
Water	4.50
	100.00

APPENDIX D.

GEOLOGY.

NOTES UPON THE SPECIMENS OF ROCKS AND MINERALS COLLECTED: BY PRESIDENT EDWARD HITCHCOCK.

REMARKS UPON THE GENERAL GEOLOGY OF THE COUNTRY TRAVERSED: BY GEO. G. SHUMARD, M. D.

APPENDIX D.

GEOLOGY.

NOTES UPON THE SPECIMENS OF ROCKS AND MINERALS COLLECTED: BY EDWARD HITCHCOCK, PRESIDENT OF AMHERST COLLEGE.

DEAR SIR: I have done what I could with the specimens you put into my hands from the Red river; but I must confess, that while these specimens, with the sections and notes by Dr. Shumard and yourself, have disclosed some interesting and valuable substances, I have found it impossible to solve several questions of importance for the want of more specimens, especially fossils. Without these, you are aware, the tertiary and secondary formations cannot be identified with any degree of certainty. Yet the whole number of species sent me does not exceed half a dozen, and several of these are so mutilated that their specific character cannot be determined. The two most important formations pointed out in your notes, and in the sections, are the gypsum deposite and that of coal; yet from the former there is not in the collection more than one species of fossil, and from the latter no specimen whatever; so that the exact place in the geological scale of these two formations is in a great measure conjectural.*

But notwithstanding these deficiencies, we do get from the specimens, and your notes, glimpses of several very valuable facts. The four most important points in your discoveries are gypsum, copper, gold, and coal. Perhaps I cannot bring out my views upon these and other points better than by describing the specimens in the order of your march, except where that was doubled upon itself. Where I can do it, and think it of any service, I shall designate by colors, upon the map of your route which you placed in my hands, the most important deposites.

At your starting point, Fort Belknap, on the Brazos river, you mention a fact of the deepest interest, viz: the occurrence of "large beds of bituminous coal." Dr. Shumard has given the following section of the strata at this place:

---

* When I wrote the above I was not aware that Dr. Geo. G. Shumard was requested to report upon the palæontology of the exploration. When that report appears, probably he, or others, can draw more accurate conclusions upon some points than I have done.

1. Sub-soil, arenaceous, and of a red color, three to ten feet.
2. Black shale, soft, and rapidly disintegrating, four feet.
3. Seams of bituminous coal, two to four feet.
4. Fine-grained sandstone, yellowish gray, with fossil ferns; thickness variable.
5. Gray non-fossiliferous limestone, of unknown thickness.

Dr. Shumard says that the fossil ferns in this formation belong to "the carboniferous era." He also describes the same formation on the third day's march, some fifty miles northeast of Fort Belknap, on one of the sources of Trinity river. He describes sandstone for several subsequent days, some of it coarse and highly ferruginous, with ripple-marks, which I should suppose might belong to the same coal measures, did he not mention that strata of red loam, so abundant in all that region, lie beneath the sandstone; which could not be, if the coal belongs to the carboniferous period. Yet he mentions that the same formation as that around Fort Belknap is largely developed between Fort Washita and Fort Smith, on Arkansas river. The latter fort is not less than three hundred and fifty miles northeast of Fort Belknap. On the 3d of May he describes "large quantities of ironstone strewn over the surface," another accompaniment of the true coal.

Now, at first view it would seem almost certain that we have here a description of a genuine coal formation of the carboniferous period, not less than three hundred and fifty miles long, associated, moreover, with those valuable iron ores which in other parts of the world are connected with such deposites; for, in descending through the formation, we find, first, overlying shale, then coal, then coal sandstone, or perhaps millstone grit, and then perhaps carboniferous limestone. But it is well known that coal occurs in other rocks besides the carboniferous, as in Eastern Virginia in oolitic sandstone, and in other places in tertiary strata. These more recent coals are often of great value, as in Virginia; but they are not generally as good as those from the carboniferous strata. It becomes an important question, therefore, to determine to what geological period the coal under consideration belongs. A few specimens of the fossil ferns would decide the matter, and I trust that Dr. Shumard is right in referring them to the carboniferous era; but it is known that analogous species occur in the higher rocks; and so, coal, even in the tertiary strata, is sometimes more or less bituminous. The evidence, however, appears to me to be strong in favor of this deposite being of the carboniferous age. But in your letter of April 1st, you state some facts respecting this coal that have thrown a little doubt over my mind You say that—

APPENDIX D.—GEOLOGY.

"The coal formation at the Brazos is found in a coarse, dark sandstone rock. which is a solid stratum, but is easily removed in consequence of being so soft. In excavating for a well, we passed through the sandstone and the coal. The greater part of the stone was removed with the mattock; and in the coal, which was here about sixty feet below the surface, we found fossil ferns, which, unfortunately, were not preserved."

The ease with which this sandstone was removed, requiring only a mattock, corresponds better with the hardness of tertiary than of carboniferous rocks; yet, in some parts of the world, distant from igneous rocks, the sedimentary strata are but little indurated.

Your statement respecting the coal on the Brazos, and the importance of the substance to the future inhabitants of the western side of the Mississippi valley, led me to recur to the journals of other explorers, as well as your own from Fort Smith to Santa Fé, published by the government in 1850, to ascertain whether this valuable mineral does not occur in such places as to justify the inference that a large coal field may exist in that portion of our country. I have not all of the necessary works of reference at hand; but, in such as I have, I have found the following cases, including those already described:

1. Fort Belknap, on the Brazos river, latitude $33\frac{1}{4}°$ to $33\frac{3}{4}°$, longitude $98°$ to $99°$.

2. Between Forts Washita and Smith, latitude $34°$ to $35\frac{1}{4}°$, longitude $94\frac{1}{2}°$ to $96\frac{3}{4}°$.

3. On Coal creek, near the South Fork of the Canadian, eighty-eight miles from Fort Smith, in longitude $96\frac{1}{4}°$, latitude $34\frac{3}{4}°$. "Bituminous coal, used by the blacksmiths of the country, who pronounce it of an excellent quality." (See Captain Marcy's report, p. 173.)

4. North branch of Platte river, latitude $42°$ to $43°$, longitude $104°$ to $107°$; described by Rev. Samuel Parker, Exploring Tour, p. 73. He calls this coal "anthracite, the same, to all appearances, as he had seen in the coal basins of Pennsylvania."

5. On the same route Colonel Fremont found coal and fossil plants in latitude $41\frac{1}{2}°$, and longitude $111°$. The fossils greatly resembled those of the true coal measures. He also found what was probably *brown* or tertiary coal, in longitude $107°$.

6. Major Emory met with "bituminous coal in abundance," in latitude $41°$, longitude $105°$. He was told of a bed thirty feet thick.

7. Lieutenant J. H Simpson describes bituminous coal in beds from two to three feet thick, in latitude $36° 12'$, and longitude $108° 52'$; and he states it to be "coextensive with the country between the valley

of the Rio Puerco and the east base of the Sierra de Tunechá, or through a longitudinal interval of $7\frac{3}{4}°$." (Report, p. 147.)

8. Lieutenant Abert found strata, which he regarded "indubitable proof of the existence of coal," in latitude $36\frac{1}{4}°$, and longitude $104\frac{1}{2}°$. (Report, p. 21.)

9. In 1818, Mr. Bringier described "a large body of blind coal (anthracite) equal in quality to the Kilkenny coal, and by far the best he had seen in the United States, immediately on the bank of the Arkansas, a little above the Pine bayou, five hundred miles from its mouth, in latitude 38°, and longitude 98°. (American Journal of Science, vol. 3. p. 41.)

10. On Monk's map of the United States, (1853,) I find two spots in Texas marked as "beds of coal," one in latitude 29°, and longitude 100°; the other in latitude $28\frac{3}{4}°$, and longitude 101°.

I might, perhaps, add that Dr. F. Roemer describes a belt of granitic and palæozoic formations, the latter of carboniferous limestone and silurian rocks, surrounded by a vast deposite of cretaceous rocks, between the Pedernales and San Saba rivers, in the northwest part of Texas. The occurrence of such rocks, especially of the carboniferous limestone, affords a strong presumption that the formation that usually lies next above this rock exists in that region.

If, now, leaving out the cases described by Fremont as most probably brown or tertiary coal, we locate the others mentioned above upon a map of the United States, we shall find a region lying between latitude $28\frac{3}{4}°$ and 43°, and between longitude $94\frac{1}{2}°$ and 109°, containing not less than nine deposites of coal, either bituminous or anthracitic; some of them one or two hundred miles long. Its northern limit is the north branch of the Platte river; its eastern limit Fort Smith, on the Arkansas; its western limit, in the country of the Navajoes, in New Mexico, and even beyond the summit level of the Rocky mountains; and its southwestern limit the Rio Grande, in the southwest part of Texas. These limits would give a north and south diameter of one thousand miles, and an east and west diameter of six hundred and eighty miles; an extent of surface three times larger than that of all the coal fields in the United States hitherto described, which cover only two hundred and eighteen thousand square miles. Yet, in view of all the facts, I think the geologist will be led strongly to suspect that a large part of this vast region at the southwest *may be* underlaid by coal. The larger part may be, and undoubtedly is, covered by newer deposites, especially the cretaceous and the tertiary; and doubtless the older rocks in Texas, as already described, may in some districts protrude through the coal measures. But if coal does actually exist beneath the newer rocks, it may be reached, as it has been in like instances in Europe, although no trace of it exists at the surface.

The above suggestions may seem to embrace a very wide field for a coal deposit. But on locating the several patches of coal upon a map of the United States, I was struck with one fact. Starting with the beds marked upon Monk,s map, in the southwest part of Texas, and running the eye along the range of carboniferous limestone described by Dr. Roemer, we come to the coal at Fort Belknap; next to the extensive deposite lying between Forts Washita and Smith, in the west part of Arkansas; and all the way we find ourselves almost in the range of the great coal field of Iowa and Missouri, as mapped by Dr. Owen; and it seems to me that every geologist will at once infer that the Missouri field does follow this line, not only across Arkansas, but also through the Choctaw Nation, and probably across Texas—interrupted, probably, in many places, by the protrusion of older rocks, and in others covered by newer formations. I have a considerable degree of confidence that such will ere long be found to be the fact, even if we leave out the other coal deposites farther west and northwest. And should the result of your explorations be to bring out such a development, I think you must feel rewarded for your fatigues and privations.

That some of the cases above described may turn out to be tertiary coal is quite possible, especially those along the base of the Rocky mountains; for it is well known that much farther to the north such coal is developed on a large scale, especially along Mackenzie's river, even to its mouth, on the Arctic ocean. Nor is it always easy for those not practised mineralogists to distinguish this coal, especially from anthracite. Dr. Owen describes the southernmost bed of brown coal on the Missouri (from four to six feet thick) as having "the aspect of ordinary bituminous coal," yet as "smouldering away, more like anthracite." (Report, p. 196.) Even such coal might be of great value; but I cannot believe that much of that described above, especially that on the line above indicated, will prove to be tertiary coal.

I ought to have mentioned, that among the specimens in my hands is one of lignite, collected July 3, near the sources of Red river, not far from the "Llano estacado," and within the limits of the gypsum deposite to be described. It is an exceedingly compact coal, and burns without flame, emitting a pungent but not bituminous odor. It is doubtless tertiary or cretaceous; but I think, if in large masses, it might easily be mistaken for anthracite.

From the 3d of May to June 2d, the formation passed over is, as I judge from Dr. Shumard's sections and descriptions, the predominant one along the upper part of Red river. All the appended sections of Dr. Shumard, except Nos. VI and XI, exhibit the characters and varieties of this deposite. Red clay is the most striking and abundant member; and above this we have a yellow or lighter colored sandstone, often finely laminated. As subordinate members, we have blue and yellow clay, gypsum, non-fossiliferous limestone, conglomerate, and copper ore. Overlying these strata is what Dr. Shumard calls "drift," which is surmounted by soil. Excepting the gypsum and the copper, no specimen of this formation was put into my hands; and only one petrefaction, which is a coral from the base of section No. IV, unless the fossil-wood belongs to it.

Now the question is, shall we regard this formation as tertiary, or cretaceous? With the means in my hands I feel unable to decide this question. If I am right in referring the fossil coral found in it to the genus *Scyphia*, as described by Goldfuss, (Petrefacta Germaniæ, Tab. XXXII, fig. 8,) it most probably belongs to the cretaceous period; for, of the one hundred and twenty species of this genus enumerated in Bronn's Index Paleontologicus, only one is found above the chalk. As to the fossil-wood, which I shall notice more particularly further on, it is well known to occur in almost all the fossiliferous deposites. Upon the whole, I rather lean to the opinion that these strata may belong to the cretaceous formation; though it is singular, if such be the case, that the fossil remains are so scarce, since, as we shall see, they occur abundantly in another portion of the field in which the cretaceous rocks abound.

Under these circumstances I shall speak of this deposite under the name of the Red Clay Formation, save where gypsum is very abundant, and then I call it the Gypsum Formation; and thus have I marked these rocks on your map.

The sandstone which constitutes the upper part of this formation has a slight dip, in a few places, of $2°$ or $3°$. On the 8th of June, however, a grayish yellow sandstone is described as having a westerly dip of $40°$; and on the 9th of June, "an outcrop of finely laminated, red, ferruginous sandstone" is mentioned, having an irregular northeasterly dip of $30°$, as shown on section VI. The next day the strata were found standing nearly perpendicular; but whether this sandstone is the same as that lying above the red clay, is not mentioned. If it is, its great dip probably results from some local disturbance. If it is not, it is probably a protruding mass of older rock exposed by denudation or upheaval.

APPENDIX D.—GEOLOGY.

The branches of Red river have cut deep chasms in this formation. In some places they are spoken of as fifty, and in others as two hundred feet deep. This clay, worn away by the streams, and mechanically suspended, gives that red color to the water, from which, without doubt, was derived the name of Red river. As to the substances held in solution by the waters of that river, some further description will be desirable before mentioning them.

The red clay formation above described abuts against the Witchita mountains, occupying the lower and more level regions around their base. Here we have an outburst of unstratified rocks, which are satisfactorily represented in the specimens.

If the relative position of the red clay and sandstone on section XI is correctly shown, I should infer some disturbance in the stratified deposites, which would indicate a more recent upheaval of the mountains than might be inferred from the nature of the rocks. The principal one is a red granite, with a great predominance of feldspar, and the almost total absence of mica. Porphyry also occurs in great quantity, of a reddish color, the imbedded crystals, for the most part, being red feldspar. In the easterly part of these mountains this rock is developed on a large scale, forming smooth, rounded hills, which slope gradually down to the plain. Cache creek passes through one of these hills, forming a gorge from three hundred to four hundred feet high, with "smooth, perpendicular walls." This rock Dr. Shumard calls *porphyritic greenstone*, and one of these walls is shown on section XI. He says that the rock is slightly columnar.

The rocks of these mountains are traversed by veins of greenstone and quartz. The latter is often porous and colored by the oxide of iron. The greenstone is the most recent of the unstratified rocks among my specimens, save a single vesicular mass, broken probably from a boulder, which has all the external marks of lava. It looks more like recent lava than any specimens I have ever met among greenstone or basalt. It was collected June 15th, west of the great gypsum deposite, though in a region abounding with sandstone, and near the bluffs that form the border of the "Llano estacado." Dr. Shumard found in the bed of the Red river, near the same place, what he calls greenstone, greenstone porphyry, and trachite. The specimen to which I have referred is rather augitic than trachitic. He says, also, that he found there "black scoria, and several other specimens of volcanic rocks." Again, on approaching the Witchita mountains on the return trip, he describes one as "a truncated cone, with a basin-shaped depression in the summit." Of this he seems to have judged by looking at the mountain from a distance. But taking all the facts into the account, I cannot but feel that there is reason to presume that volcanic agency has been active in that region more recently than the trap dykes.

I ought to add, that before reaching the Witchita mountains Dr. Shumard met with large quantities of dark-colored and cellular igneous rock, composed principally of silex and carbonate of lime, strewed over the surface. This was on the 18th of May, and on the 27th he "frequently encountered local deposites of red, scoriaceous rock." Among the specimens in my hands are some apparently more or less melted, composed of carbonate of lime and copper ore.

Again, scattered widely over the surface, numerous specimens were found of jasper, carnelian, and agate. The carnelian is deep red, but found in botryoidal, or even stalactitical masses, and they have seemed to me to resemble more those silicious nodules found in soft limestone than in trap rocks. They were found most abundantly towards the western part of the region gone over.

I ought to have mentioned that the Witchita mountains consist of numerous peaks, rising from eight hundred to nine hundred feet above the river. Mount Webster, one of the most conspicuous, was found to be 783 feet above the plain by the barometer. Twelve of these elevations were found to be composed of granite, which in many places is undergoing rapid disintegration.

We have seen in the red clay of this region a reason for the name of Red river, and the character of its waters. In the above description of the rocks of the Witchita mountains, I think we may see the origin of the red clay. The great amount of iron which they contain would produce exactly such a deposite upon their decomposition and erosion by water. And we have reason for supposing this red granite to be a quite extensive formation, as I shall shortly show.

No one at all acquainted with the rocks in which gold is found can look at the specimens you have obtained in the Witchita mountains without expecting that he shall be able to detect that metal. The porphyry, the porous quartz from veins impregnated with hydrate of iron, and the magnetic iron-sand found in the bed of Otter and Cache creeks, excite this expectation. In one of your letters you state that "the people of Texas have for a long time supposed that there was gold in the Witchita mountains, and they have attempted to make several examinations for the purpose of ascertaining the fact, but have invariably been driven away by the Indians. We searched diligently about the mountains, but could find only two very minute pieces imbedded in quartz pebbles." This, as Dr. Shumard states, was upon Otter creek, and there occurred the ferruginous sand, which occurs also upon Cache creek in great quantities. We have not been so fortunate as to find any gold in the specimens sent, although the sand has been carefully examined, and two assays have been made of the quartz in the laboratory. Yet I can easily believe that gold must exist either among that black sand, or in the veins of ferruginous quartz—sometimes three feet wide—so common in the Witchita mountains.

It is well known that a good deal of excitement exists on this subject at the present moment in Texas; but the "gold diggings" there lie upon the upper Colorado. From some able remarks on the subject in the "Telegraph and Texan Register" of April 29th, by the editor, Francis Moore, jr., I learn that the region where the gold is found is "a belt of fifteen or twenty miles wide, which extends from the sources of the Gaudalupe, by the Enchanted Rock, to the head of Cherokee creek, a branch of the San Saba." The description of that belt which follows, as you will see, corresponds very well to the region around the Witchita mountains. "The red granite rocks here crop out above the secondary formations, and veins of quartz are found traversing the rocks in all directions. The soil is generally of a red mulatto color, caused by the decomposition of the red feldspar of the granite. These rocks resemble, it is said, those of the gold regions of California and Santa Fé. A gentleman who has recently visited the Nueces states that gold has also been found on that river; and if the report that gold has been found in the Witchita mountains be correct, it is possible that this narrow belt of primitive rocks extends quite through from the Nueces to those mountains, a distance of about four hundred miles. It is mentioned in Long's Expedition that a narrow belt of red granite is found jutting up through the prairie region on the Des Moines river, in Iowa, and it is not improbable that this is a continuation of the primitive ridge, extending by the Witchita mountains and the Enchanted Rock, to the sources of the Nueces, and it may extend far above Lake Superior." As to this northern extension of these gold-bearing rocks, I do not find much to confirm the conjecture in Dr. Owen's late able report on that region, although he does mention some red granite and some red clay; but the latter is probably alluvial. Yet, that these rocks may extend through Texas, and even much farther north, is extremely probable.

But though your discovery of gold will probably excite more attention, I feel that the great gypsum deposite of the West, which you have brought to light, will be of far more consequence to the country.

On your map I have colored this formation as you have marked it out. Yet I cannot doubt, from the descriptions and sections, that the gypsum is embraced in the red clay formation already described, for most of this mineral occurs above the red clay, though sometimes embraced within it. Yet the importance of the gypsum justifies me in coloring that portion of these strata as the gypsum formation where it is most abundant. It is several times mentioned as occurring in other parts of the region, marked as red clay. But on the 3d of June, high bluffs were met of red and blue clay, with interstratified layers of snow-white gypsum. From this time till the 12th the same formation was found, and also from the 21st of June to the 9th of July. But your own description of this formation in your letter of November, 1852, contains a better account of its extent than I can give.

"I have traced this gypsum belt," you observe, "from the Canadian river, in a southwest direction, to near the Rio Grande, in New Mexico. It is about fifty miles wide upon the Canadian, and is embraced within the 99th and 100th degrees of west longitude. Upon the North, Middle, and South forks of Red river it is found, and upon the latter is about one hundred miles wide, and embraced within the 101st and 103d degrees of longitude. I also met with the same formation upon the Brazos river, as also upon the Colorado and Pecos rivers, but did not ascertain its width. The point where I struck it, upon the Pecos, was in longitude $104\frac{1}{2}°$ W.

"Wherever I have met with this gypsum I have observed all the varieties from common plaster of Paris to pure selenite; and among specimens of the latter were pieces *three feet by four, two inches in thickness*, and as perfectly transparent as any crown glass I have ever seen. It is to be regretted that I could not have brought home some of these beautiful specimens; but my means of transportation were too limited. I regard this gypsum belt as a very prominent and striking feature in the geology of that country. From its uniformity and extent I do not think there is a more perfect and beautiful formation of the kind known. I have myself traced it about three hundred and fifty miles, and it probably extends much further."

The position and thickness of the gypsum beds may be learnt from Dr. Shumard's sections, especially No. V, where they are from ten to fifteen feet thick. I do not wonder that you have been deeply impressed with the vast extent of this deposite. Prof. D. D. Owen, in his late valuable report of a geological survey of Wisconsin, Iowa, &c., (1852) describes a gypseous deposite, twenty to thirty feet thick, in the carboniferous strata, and occupying an area from two to three square miles; and he says, that "for thickness and extent, this is by far the most important bed of plaster-stone known west of the Appalachian chain, if not in the United States." (p. 126.) Either deposite may be large enough to supply the wants of the inhabitants who may live near enough to obtain it. But the vast extent of your deposite (doubtless greater, as you say, than is at present known) will make it accessible to much the greatest number of people. Indeed, from the well known use of this substance in agriculture, as well as other arts, a knowledge of its existence must have an important bearing upon the settlement and population of northwestern Texas.

The only deposites of gypsum known to me that are more extensive than the one discovered by you, are in South America. All along the western side of the Cordilleras, especially in Chili, and interstratified with red sandstone and calcareous slate, beds of gypsum occur of enormous thickness, some of them not less than six thousand feet. It has been tilted up and metamorphosed greatly by igneous agency of ancient date, but seems to be of the age of the lower cretaceous rocks. Mr. Darwin, to whose admirable work on the geology of South America I am indebted for these facts, has traced this deposite at least five hundred miles from north to south, (it is not many miles—sometimes, however, twenty or thirty—in width,) and thinks it extends five hundred more; and perhaps much further. He also describes thin beds of gypsum in the tertiary strata of Patagonia and Chili, which are some eleven hundred miles in extent. This gypsum is generally more or less crystalline, and corresponds much better in lithological characters with that in Texas, than does the metamorphic gypsum of the Cordilleras. Mr. Darwin is of opinion, however, that the latter was originally deposited in a manner analogous to the former, viz: by means of submarine volcanoes and the conjoint action of the ocean. Very probably the ancient igneous agency which we have described in the Witchita mountains, and along a line southerly to the Rio Grande, may have been connected with the production of the gypseous deposite in the same region.

The specimens of this gypsum put into my hands correspond with your descriptions. One of them, of snowy whiteness and compact, it seems to me, might answer for delicate gypseous alabaster, so extensively wrought in other lands for ornamental purposes. The selenite was regarded among the ancients as the most delicate variety of alabaster, and was employed by the wealthy, and in palaces, for windows, under the name of *Phengites*. It has the curious property of enabling a person within the house to see all that passes abroad, while those abroad cannot see what is passing within. Hence Nero employed it in his palace. If the splendid plates which you describe occur in any considerable quantity, it may hereafter be of commercial value, as it certainly will be of mineralogical interest.

From your description, especially in your lecture before the American Geographical and Statistical Society, it is manifest that the character of the rocks changes on the northwest of the gypsum formation, and near the head of the south branch of Red river. The red clay and gypsum have disappeared, and sandstone succeeds; but of what age I have no means of judging.

Another interesting mineral found by you in the red clay and gypsum formations above described, is copper. The specimens were put into the hands of Professor Charles U. Shepard, who has analyzed them, as well as several other specimens, in the laboratory of Amherst College, and whose report I annex to my own. You will see that he has made free use of your name by attaching it to a new ore of copper, found on Red river near the Witchita mountains; and that he describes three or four other species of copper ore from the same region. For a particular description I refer you to his report, while I confine myself to a few remarks as to the geology of the deposite.

On section V, Dr. Shumard has shown the geological position of this ore, viz: near the bottom of, and in the red clay, and more than one hundred feet from the surface. We hence see that the ore was deposited from water, although some specimens from Cache creek of calcareous amygdaloid seem to have been melted. But if, as has been suggested, the gypsum was produced by the joint action of submarine volcanoes and water, the copper may have had the same origin, and this would explain the presence of chlorine in the *Marcylite*.

How much copper may be expected in such a region as that on Red river, I have no means of judging, because I know of no analogous formation. But as we have proof that it is an aqueous deposite, and that igneous agency has been active not far off, it would not be strange if the vicinity of Witchita mountains should prove a prolific locality.

The oxide of manganese described by Professor Shepard may, perhaps, be found abundant and more pure. And the iron sand, so common in some of the creeks, indicates the existence of magnetic oxide of iron in the mountains.

Whether the red clay formation and the gypsum formation that have been described are of the cretaceous age or not, there can be no doubt as to the deposites passed over from July 20 to Fort Washita, for among the specimens are two species of *Gryphœa*, and one echinoderm, much mutilated, but evidently of that period. On the 27th, a "bluish gray, highly crystallized limestone" was observed, which cropped out beneath the sandstone, and which Dr. Shumard says was "non-fossiliferous." It continued, however, to the 30th, or to Fort Washita, where he says, " I observed in it a large number of the fossils characteristic of the cretaceous period." Probably he refers to two kinds of limestone, and not improbably the limestone and sandstone first noticed belong to the carboniferous strata already noticed. Among the specimens I also find parts of two species of ammonite; one quite large, but quite characteristic of the cretaceous strata, and resembling some good specimens in the collection of the American Board of Foreign Missions, obtained by their missionaries in the Choctaw country. I cannot doubt that these strata are largely developed in that vicinity. Indeed, that region has already been colored as of the cretaceous age upon our geological maps. I have, therefore, marked a strip of cretaceous rocks between Forts Belknap and Washita. These are, in truth, the predominant strata in Alabama, Mississippi, and Texas, and I need not go into details respecting them.

Dr. Shumard frequently speaks of a surface formation under the name of *drift*, consisting of boulders of all the rocks described above, and some others, such as mica slate and labradorite. But I doubt whether this formation be the same which we denominate drift in New England—the joint result of water and ice; for no example has as yet been found of drift agency as far south as Texas, by several degrees. Yet there is evidence of a southerly movement among the smaller rolled detritus almost to the Gulf of Mexico, such as water alone could produce, seeming to be the result of the same current, destitute of ice, that produced the coarse unstratified and unsorted drift of Canada and New England. But among the specimens in my hands are several of silicified wood, and all of them, I believe, are mentioned in Dr. Shumard's notes as occurring in drift; although in your letter of December 5, 1852, you speak of masses from fifty to one hundred pounds in weight in the gypsum formation. You may mean in its upper part:* if so, there may be no discrepancy between the two statements; and I have been led to suspect that what Dr. Shumard calls drift may be only a newer portion of the tertiary strata, although, as already remarked, silicified wood is found in almost all the fossiliferous formations. All the specimens sent by you, however, with one exception, are dicotyledonous. They resemble not a little the fossil-wood from Antigua, and the desert near Cairo, in Egypt; both of which deposites are tertiary. One specimen is a beautiful example of a monocotyledon, a cross sec-

---

*The fossil-wood referred to in Captain Marcy's letter was found upon the upper surface of the formation.

tion showing vessels of the shape of a half or gibbous moon. This fact shows that the climate was warm enough for trees analogous to the palm tribe to flourish; yet the great predominance of dicotyledonous forms shows a close analogy with the existing vegetation of the southern part of our country; nor is there evidence, in these specimens, of a temperature above that now existing in our southern States, since several species of palms occur there.

The two subsoils analyzed by Professor Shepard, give very interesting results. The first is highly calcareous; and when the lime shall have been exhausted in the overlying soil, this material, thrown up by subsoil ploughing, would be equal to a large dressing of lime. In the other subsoil we have an extraordinary amount of sulphate of lime, and a sufficiency of carbonate of lime, as well as chlorine and soda. It seems hardly possible to doubt that such a basis would need only organic matter to render it one of the most productive of all soils; and when we think how extensive the gypsum formation is from which this subsoil was obtained, we cannot but anticipate (unless there are counteracting causes of which I am ignorant) that that portion of our country will become a rich agricultural district—I mean the region lying east of the "Llano estacado."

Only one specimen of common salt (chloride of sodium) was sent among the specimens, and that, as you inform me, "was procured by the Comanche Indians in the country lying between the Canadian and Arkansas rivers."

We are now prepared to appreciate an analysis of the water of Red river, which has been executed in the laboratory of Amherst College by Mr. Daniel Putnam, under the direction of Professor W. S. Clark. This is somewhat of a mineral water, and you remark that all the waters originating in the gypsum formation have the same bitter and nauseating taste. I think you are right in the opinion that the ingredients are derived from that formation. Analysis shows that the taste depends upon the presence of three salts in nearly equal proportions, two of which, sulphate of magnesia, or Epsom salts, and chloride of sodium, are very sapid. Mr. Putnam's analysis is as follows:

"Water from Red river—

Water in fluid ounces	4
Weight of water in grammes	127.800
Weight of chlorine present	.051
Weight of lime present	.033
Weight of sulphuric acid present	.095
Residue evaporated to dryness, and weighed, *probably*, sulphates of soda and magnesia together, weight	.168

"It was impossible, with the small quantity of water, to determine the last two ingredients with *absolute certainty*. In the calculations following they are regarded as *real*. Regarding the lime as sulphate, and the residue of sulphuric acid as united with magnesia, and the chlorine as united with sodium, we have the following results:

Weight of sulphate of lime	.080
Weight of sulphate of magnesia	.073
Weight of chloride of sodium	.084
Weight of the whole	.237

Per-centage of matter in solution, about	.19

"The analysis of the water from a spring in a gypsum cave, yielded the following results:

Weight of the water, in fluid ounces	4
Weight of the water in grammes, about	127.800
Weight of hydro-sulphuric acid present	.011
Weight of chlorine	.014
Weight of lime	.090
Weight of sulphuric acid	.227

The residue was evaporated, and the *presence*, but not the weight, of magnesia, found separate from the soda. The quantity was very small, however.

Soda and magnesia together, about	.130

"Regarding the lime as sulphate, and the residue of sulphuric acid as united partly with magnesia and partly with soda, and the chlorine with sodium, we have the following results:

Weight of sulphate of lime	.219
Weight of sulphate of magnesia	.088(?)
Weight of sulphate of soda	.073(?)
Weight of chloride of sodium	.023
Weight of hydro-sulphuric acid	.011
Weight of the whole	.414

Per-centage of matter in solution	.82

"The analyses of water, on account of the small quantity, cannot be relied upon as perfectly accurate; but they are the best I could make under the circumstances."

Your account of the remarkable *cañons* of Red river, where it comes out from the borders of the "Llano estacado," as given in your lecture before the American Geographical and Statistical Society, has been read by me with great interest. For several years past I have been engaged in studying analogous phenomena in this, which seems to me a neglected part of geology. The cañons of our southwestern regions are among the most remarkable examples of erosions on the globe; and the one on Red river seems to me to be on a more gigantic scale than any of which I have found a description. You seem in doubt whether this gorge was worn away by the river, or is the result of some paroxysmal convulsion. You will allow me to say that I have scarcely any doubt that the stream itself has done the work. The fact that when a tributary stream enters the main river it passes through a tributary cañon, seems to me to show conclusively that these gorges were produced by erosion, and not by fractures; for, how strange would it be if fractures should take those ramifications and curvatures which a river and its tributaries present. And, moreover, I find cases where I can prove, from other considerations, that streams of water (existing and ancient rivers) have eaten out gorges quite as difficult to excavate as any of the cañons of the West. So that, if we must admit that rivers have done a work equally great in one case, all presumption is removed against their doing the same in other cases. I have a great number of facts which I hope to be able, if life be spared, to present to the public on this subject; and I am very glad to add the cañons of Red river to the number.

Before Professor Adams's departure for the West Indies last winter, I secured his report, hereto subjoined, upon the recent shells collected in your expedition. It derives a melancholy interest from having been among the last, if not the very last, of his scientific efforts, he having been cut off by yellow fever in January.

With this imperfect elucidation of the facts collected by you in your laborious explorations, I subscribe myself,

With great respect,
Your obedient servant,
EDWARD HITCHCOCK.

AMHERST COLLEGE, *June* 5, 1853.

REMARKS UPON THE GENERAL GEOLOGY OF THE COUNTRY PASSED OVER BY THE EXPLORING EXPEDITION TO THE SOURCES OF RED RIVER, UNDER COMMAND OF CAPTAIN R. B. MARCY, U. S. A.: BY GEO. G. SHUMARD, M. D.

It is to be regretted that the main objects contemplated by the expedition were of such a character as to allow of merely a partial geological exploration. It was found necessary to traverse a large extent of country in a limited period of time, so that not as many opportunities were allowed for making minute and detailed sections of the strata as could have been desired. However, it is believed that something has been done towards elucidating the geology of a valuable and interesting district of our country, which hitherto has received but little attention from geologists.

We will first submit a brief account of the geological features of a portion of Northwestern Arkansas, which will enable us to understand more clearly the character of the deposites observed on the route travelled by the party, and exhibit more satisfactorily the connection of the cretaceous group with the older or palæozoic rocks. In Washington county we have a fine development of rocks belonging to the carboniferous period, rising sometimes several hundred feet above the water-level of Arkansas river. They consist of beds of dark-gray and bluish-gray limestone, surmounted by heavy-bedded coarse and fine-grained quartzose sandstone. The ridges of highest elevation run nearly north and south through the centre of the country, forming a geological back-bone; the waters from one side flowing eastwardly into White river, and on the other westwardly into Illinois river, both streams being tributaries of the Arkansas.

Wherever the limestone forms the surface-rock, the soil is of excellent character, and for productiveness is unsurpassed by any in the State; but where the sandstone reaches the surface, the soil becomes too arenaceous, and is of inferior quality for agricultural purposes. The limestone is generally highly charged with fossils, and, in many places, beds of considerable thickness are almost entirely composed of the remains of Crinoidea.

In lithological and palæontological characters it corresponds very closely to the rocks of the superior division of the carboniferous system of Indiana, Kentucky, Illinois, and Missouri. The fossils are usually remarkably well preserved. The following are the most abundant and characteristic species: *Archimedipora archimedes, Agassizocrinus dactyliformis, Pentatematites sulcatus, Productus cora, P. punctatus, P. costatus, Terebratula subtilita,* and *Terebratula Marcyi.** We have found all these species associated together in Grayson county, Kentucky, near Salem, Indiana, and at Chester and Kaskaskia, Illinois.

The line of junction between the sandstone and limestone is well defined, there being an abrupt transition from the one into the other. The sandstone has yielded but few fossils, and these only calamites and ferns.

Veins of sulphuret of lead traverse the limestone at several points in Washington county, and I have been informed that valuable beds of iron ore occur here; workable seams of bituminous coal have also been discovered at a number of localities in the county.

Proceeding in a southerly direction through the counties of Crawford and Sebastian, the limestone, which, with few exceptions, constitutes the surface-rock in Washington county, dips beneath the sandstone, and the latter forms the entire mass of the hills, rising sometimes to, the altitude of a thousand feet above the adjacent streams: it is, for the most part, the prevailing rock the entire distance between Fort Smith and Camp Belknap. The sandstone is often highly ferruginous, and varies in color from light-gray to dark-brown. It exists in heavy massive beds, made up of coarse quartzose grains, with intercalations of finer-grained sandstone, occasionally beautifully ripple-marked. It corresponds in its lithological features with that forming the Ozark range of mountains.

In Sebastian county I found a few *Calamites, Lepidodendra,* and several varieties of fossil ferns of the coal formation, but organic remains are by no means abundant. Bituminous coal exists in almost inexhaustible quantities throughout the county. The seams vary in thickness from a few inches to seven feet, and they lie in such a manner that they can be wrought easily. Coal has also been discovered at a number of localities between Fort Smith and Fort Washita.

About a hundred miles southwest of Fort Smith we encountered an outcrop of bluish-gray limestone, which extends across the country in a southeasterly direction for the distance of about twenty miles; it presents an average thickness of about ten feet, with a dip to the east of 30°. Its precise character could not be determined, as we were unable to find any fossils.

---

* Figures and descriptions of the fossils of these beds will be found in the appended report of Dr. B. F. Shumard on the palæontology of the expedition.

APPENDIX D.—GEOLOGY. 181

Pursuing the same direction, twenty-five miles beyond is an outburst of granite, which extends for the distance of twenty-six miles, with a southerly bearing. This is the only example of rocks of igneous origin to be met with between Forts Smith and Preston, and the rough and rugged features of the country where it prevails, forms a striking contrast with the comparatively rounded outline of sandstone hills. The rock is of a coarse texture, and varies in compactness in different portions of the range; feldspar of the flesh-colored varieties predominates over the other ingredients. In places the rocks would form an excellent and durable building material, but in other portions of the range it crumbles readily when exposed to the action of the weather.

We observed numerous veins of quartz traversing the granite in various directions, and, at some points, dykes of compact greenstone porphyry. Saline springs were found not unfrequently issuing from the base of the range, and the waters in one or two instances were found so strongly impregnated with saline matter, as to induce the belief that they might be worked with profit.

Passing this range, the sandstone again reappears, and constitutes the prevailing rock to within a short distance of Fort Washita, where it disappears, and is succeeded by strata of the cretaceous period.

From this point the cretaceous rocks were found to extend uninterruptedly until we reached the southwestern boundary of the Cross-Timbers, in Texas. From the best information I was able to procure, it constitutes the prevailing formation from Fort Washita in the direction of Fort Towson for upwards of a hundred miles, with an average breadth of fifty miles. It forms part of that extensive belt of cretaceous strata that extends from Georgia to Texas, and which, from the character of its fossil fauna, is now regarded as the equivalent of the upper chalk of England, and with that division of the cretaceous group to which D'Orbigny gives the name of *l'Etage Senonien*, (Prodrome de Palæontologie, tome II, page 669.) Wherever sections of the strata were to be seen they presented the following characters: grayish-yellow sandstone, with intercalations of blue, yellow, and ash-colored clays, and beds of white and bluish-white limestone. The limestone reposes on the clays and sandstones. At some points it attains the thickness of a hundred feet; while at others it is quite thin, and sometimes even entirely wanting. It is usually soft and friable, and liable to disintegrate rapidly when exposed to the action of the weather. These cretaceous rocks are often full of fossils. At Fort Washita the layers are crowded with *Ananchytes, Hemiaster, Nucleolites, Ammonites, Ostrea, Pecten*, &c., descriptions and figures of which will be found in Dr. B. F. Shumard's report on the palæontology of the expedition. We saw here some specimens of ammonites several feet in diameter, and weighing between four and five hundred pounds. On Red river, twenty-six miles from Fort Washita, the sandstone of the cretaceous group supports about twenty-five feet of ash-colored calcareous loam, which, on inspection, was found to contain terrestrial and fluviatile shells of the genera *Lymnea*, *Physa*, *Planorbis*, *Pupa*, and *Helix*, the whole resembling species which we have observed in the loam at New Harmony, Indiana, and elsewhere in the Mississippi Valley, which Mr. Lyell, during his visit to this country, recognised as the equivalent of the loess of the Rhine.

The geological formation, as developed in the vicinity of Camp Belknap, consists of nearly horizontal strata of fine-grained sandstone, shale, and soft, drab-colored, non-fossiliferous limestone, whose relative positions correspond with strata of the same character largely developed between Fort Washita and Fort Smith. On the surface were in many places strewn fragments of a reddish-gray, igneous rock, containing a large per-centage of carbonate and oxide of iron. From the frequent indications of the presence of that metal in various localities of this region, it is not improbable that this may become hereafter an extensive and profitable field of mining enterprise. Recently a number of seams of bituminous coal, varying in thickness from two to four feet, as well as the characteristic fossil ferns of the carboniferous era, have been discovered.

The following section, taken about one mile from the post, may give a better idea of the formation:

1. Subsoil arenaceous, and of a red color; thickness from three to ten feet.
2. Black shale, soft, and rapidly disintegrating; four feet thick.
3. Seams of bituminous coal, from two to four feet thick.
4. Fine-grained sandstone, of a yellowish gray color, and containing fossil ferns; thickness variable.
5. Gray non-fossiliferous limestone; thickness unknown.

The water obtained from springs in this vicinity frequently contains iron in solution. I have been informed that in a few instances chloride of sodium has been detected in it.

May 3.—Formation the same as at Camp Belknap. Observed, strewn over the surface, large quantities of iron-stone; soil and subsoil arenaceous, and deeply tinged with oxide of iron.

May 4.—Saw a number of horizontal layers of coarsely laminated sandstone; between the laminations were observed a large number of ripple-marks. Soil good, and of a dark color; subsoil, in some places, arenaceous, in others argillaceous, and of a deep-red color.

APPENDIX D.—GEOLOGY.

May 5.—For the first six miles the surface became gradually more elevated. Here, and elsewhere to-day, we met with a number of horizontal layers of coarse-grained and highly ferruginous sandstone, which was more or less laminated, and highly embossed with ripple-marks. In many places we met with extensive deposites of porous and dark-colored igneous rock, containing a large per-centage of oxide of iron. The surface was everywhere strewn with drift, mostly composed of quartz, greenstone porphyry, and granite. Saw a number of conical hills, varying in height from ten to seventy-five feet, and composed of horizontal layers of sandstone, of the same character as that first met with to-day. Owing to the rapid disintegration of the sandstone, the hills are gradually crumbling away. In many places we found a few loose fragments of sandstone, intermixed with sand, the only indication left of the previous existence of many of them. In this manner has a levelling process gone on for ages, which, if not interfered with, will ultimately tend to the removal of the various inequalities of the surface of the prairies. Soil good; subsoil argillaceous, and of a deep-red color: this mixing in the form of sediment with the water, imparts to it a red color and disagreeable taste. From the north branch of the Witchita I collected a number of bivalve shells of the genus *Unio*.

May 6.—Sandstone and drift the same as yesterday. Saw a number of bluff banks, varying in height from ten to fifty feet. They were composed of red loam, the relative position of which was found to be below that of the sandstone. Soil and subsoil the same as we passed yesterday.

May 7.—Formation the same. Drift appears to be gradually becoming more abundant.

May 8.—During the day we had frequent opportunities of observing the sandstone and red loam. Their relative positions were the same as before, and dipped in various directions at angles of from one to three degrees. Saw a number of small boulders, composed of granite and greenstone porphyry.

May 9.—Did not move from camp. In the afternoon I explored a few miles along the banks of the Big Witchita. The geological formation, as there developed, consisted of finely laminated, soft, ferruginous sandstone, interstratified with red clay, together with drift, which last was much coarser than any previously observed. Soil good; subsoil loamy.

May 11.—Formation the same as before. Found a number of specimens of peroxide of iron.

May 12.—Red river, as observed to-day, runs through a thick bed of red loam, which, mixing with the water, imparts to it highly characteristic red sedimentary properties. Its bed was composed of fine sand. After travelling about six miles we came to a small creek with high bluff banks, near the base of which I observed a number of specimens of green and blue copper ores. Associated with it, as a matrix, was a porous and dark-colored igneous rock, containing disseminated particles of green copper ore. At this point I had an opportunity of observing the aqueous strata, from which I obtained the following section:

1. Black argillaceous subsoil; six feet thick.
2. Soft fine-grained sandstone, of a grayish color; five feet thick.
3. Red and blue clay; from six to ten feet thick.

These strata presented an easterly dip of nearly two degrees. I saw during the day large quantities of drift and a few small boulders, composed of granite, quartz, and greenstone porphyry. In a few hours we arrived at Cache creek, which runs between high bluff banks composed of red clay; its bottom was thickly strewn with large, angular fragments of quartz, greenstone porphyry, granite, and hornblende rock. Within a short distance from the creek we found a small spring of clear water, which was strongly impregnated with sulphuretted hydrogen gas. Soil dark and fertile; subsoil argillaceous, and of a deep-red color.

May 14.—Did not move from camp. In the evening I rode to the junction of Cache creek and Red river, near which point I observed a stratum of finely laminated ferruginous sandstone; in some places it was interstratified with red clay, and presented a south-southeasterly dip of three degrees, (see Section No. 3.) Saw scattered over the surface a number of small boulders of the same composition as those of yesterday. Soil black and fertile; subsoil argillaceous.

May 15.—Did not move from camp; tested the water of Cache creek, and found it strongly alkaline. Its temperature was 75° F.

May 16.—Passed to-day a number of long, low ridges, presenting on one side a gradual slope towards the prairie-level; on the other, abrupt precipitous terminations. They were for the most part composed of dark-colored scoriaceous rock, containing a moderate per-centage of copper ore. About 8 o'clock we came to a small creek, near which I observed a deposite of soft granite, which appeared to be undergoing rapid disintegration. The banks of the creek were composed of horizontal layers of finely laminated sandstone, deeply tinged with copper, and resting upon a base of red indurated clay. Saw to-day large quantities of drift, containing small boulders, composed, as before, of greenstone porphyry, quartz, and granite; soil and subsoil arenaceous.

May 17.—Formation the same as on yesterday; saw strewn over the surface a large quantity of reddish-brown and black calcareous rock, containing carbonate of copper and small crystals of calcareous spar. From the drift (which appears to be becoming more abundant and its particles less rounded) I obtained specimens of chalcedony, jasper, and carnelian. Soil and subsoil arenaceous, and of a reddish color.

May 18.—Saw a number of deposites of soft, coarse granite, which appeared to be undergoing rapid disintegration. The surface presented large quantities of dark-colored and cellular igneous rock, composed principally of silex and carbonate of lime; soil and subsoil arenaceous.

May 20.—Observed several clear springs bubbling up from beneath the surface. Formation the same as before; soil and subsoil arenaceous.

May 21.—Met to-day with several sections of finely laminated sandstone of the same character as that before mentioned, with the exception that the different laminæ were thickly marked with small circular spots of a green and yellow color. In several places I found it interstratified with red clay. Near our encampment a fine section, showing an anticlinal axis, the strata dipping east and west at an angle of three degrees, exposed itself; over the surface were strewn large quantities of dark-colored igneous rock of the same character as that seen on the 18th instant. The drift was less abundant than before; soil and subsoil arenaceous.

May 22.—The surface was strewn in many places with detritus composed of greenstone porphyry and granite; soil and subsoil arenaceous.

May 23.—Did not move from camp; in the evening I explored Otter creek, which at this point runs between bluff banks composed of red clay. Its bed was thickly covered with drift, from which I obtained a number of agates and two small specimens of bluish-yellow quartz, each containing a small particle of gold. By digging a few inches below the drift, I reached a deposite of black ferruginous sand, which, upon being stirred, emitted a strong odor of sulphuretted hydrogen gas. From the creek I obtained a number of univalve and bivalve shells; the latter principally of the genus *Unio*.

Captain Marcy having to-day visited several of the mountains, presented me with a number of specimens of soft granite of a reddish-brown color, and of which the mountains appeared to be composed.

May 25.—Remained in camp. This afternoon I measured with a thermometer the temperature of Otter creek, and found it to be 72° F.

Immediately adjoining the creek the soil is good and very productive; but at a little distance from it, it is barren and sandy.

May 26.—To-day we passed a number of sand-hills, varying in height from ten to thirty feet. The only rocks met with were a few small boulders, composed of quartz and greenstone.

May 27.—The surface was in many places composed of detritus of granite, quartz, and greenstone; saw to-day a number of boulders, mostly composed of hard granite, and presenting smooth and polished surfaces. The largest was about fifteen feet in circumference, and would weigh probably three or four thousand pounds. We frequently encountered local deposites of red scoriaceous rock. Captain McClellan having visited one of the mountains, presented me with a specimen of gray calcareous sandstone, which, as he informed me, he obtained from a horizontal stratum of the same, situated within a few feet of the base of the mountain.

Thus far about twelve of the Witchita mountains have been examined, and have been found to present a nearly uniform appearance and structure. Composed of fine granite of various degrees of hardness and color, they rise abruptly from a smooth and nearly level plain to the height of eight or nine hundred feet. Many of them are isolated and of an irregular conical shape, while others are grouped together in small clusters, and are more or less rounded. At a distance they appeared to be smooth, but upon a nearer approach their surfaces were found to be quite rough, and presenting the appearance of loose rock thrown confusedly together. In many places the granite was observed occupying its original position, and was variously traversed by joints and master-joints, which, intersecting each other at right-angles, gave to the mass somewhat of a cuboidal structure. Soil rich, and from three to four feet thick; subsoil argillaceous and of a red color.

May 28.—Did not move from camp. In the evening I explored a short distance up and down Otter creek; its bed is here composed of horizontal layers of finely laminated sandstone, containing green and yellow spots of the same character as those noticed on the 21st instant.

May 29.—Passed a number of the mountains, several of which I ascended and found composed of hard granite, variously traversed by veins of greenstone porphyry and yellow quartz; the last containing small scales of mica. The sides of the mountains frequently presented lofty precipices, one of which was divided from top to bottom by a vein of greenstone nearly perpendicular, and about twenty inches thick. I observed no change in the character of the adjoining prairie, except a few local deposites of drift and detritus, from which I collected specimens of chalcedony, agate, and jasper. No rock of any description was observed at a greater distance than a few feet from the base of the mountains. Soil thick and fertile; subsoil loamy.

May 30.—The mountains did not differ materially in appearance or structure from those before observed; at a distance, a few of them appeared to present a columnar structure, but upon a nearer approach this was found to be owing to divisional plains, or master-joints, with weather-worn and rounded edges. I observed to-day a number of clear springs; the water of several being tasted, was found to be alkaline.

In the prairie we observed several circular elevations, varying from one hundred to one hundred and thirty yards in diameter, and ascending in some places to the height of three or four hundred feet above the general level. Upon examination, their mineralogical composition was found to be the same as that of the neighboring mountains. Within a few feet of one of these, a small ravine exposed to view a horizontal stratum of soft ferruginous sandstone. Soil and subsoil the same as on yesterday.

May 31.—The mountains presented the same general appearance as on yesterday. From their surface were exhibited a large number of veins, varying in thickness from an inch to a foot and a half, and composed of greenstone, quartz, and hornblende. The prairie was here and there dotted with a number of conoidal elevations, varying in height from twenty to one hundred feet. In composition they agreed in every respect with the neighboring mountains, with which in origin they appeared to be cotemporaneous. From the drift I collected specimens of fossil-wood. The water of springs issuing from the mountains I found, upon test, to be alkaline.

June 1.—Red river as observed to-day runs between low bluff banks, composed of red clay. Its bed was in some places thickly strewn with large detached masses of granite, all presenting a highly water-worn appearance, and seeming to have been derived from a neighboring mountain. Soil and subsoil the same as before.

June 2.—Immediately upon leaving the Witchita mountains, we lost all traces of drift and other igneous rocks. Red river as observed to-day runs between high bluff banks, composed of horizontal layers of red, yellow, and blue clay, and finely laminated sandstone; the latter being interstratified with thin seams of saccharoid gypsum, (see Section No. 4.) About a mile from the river, we observed two conical hills—one fifty and the other eighty feet in height—composed of horizontal layers of sandstone, interstratified with thin seams of gypsum. From them I obtained specimens of selenite. Soil and subsoil loamy.

June 3.—To-day we came to a range of high bluffs about six miles in length, and extending in a direction nearly parallel with the river. At a distance they resembled a long line of fortification; upon examination they were found to be composed of horizontal layers of red and blue clay, thickly interstratified with snow-white gypsum, (see Section No. 5.) These bluffs appeared to be rapidly yielding to the weather: along their base were thickly strewn large cuboidal masses of gypsum—some ten feet in diameter—that appeared to have been but recently detached from a stratum of the same near their summits. In the blue clay I observed a thin seam of carbonate of copper. The gypsum was also in a few places slightly tinged with the same metal. In a southerly direction, and at the distance of about fifteen miles, we observed another range of gypsum bluffs: they appeared to run in a direction parallel with those already described. The intervening country was very rough and broken. Soil dark and fertile; subsoil argillaceous.

June 4.—Passed a number of bluffs of the same composition as those observed yesterday. The surface during the greater portion of the march was whitened by gypsum, which was always found occupying its position above the red clay. In the evening I visited a small hill, situated about three miles from camp, and succeeded in discovering a thin seam of copper ore, as well as large beds of selenite. Soil and subsoil the same as on yesterday.

June 5.—The country travelled over to-day was mostly composed of sand-hills, varying in height from ten to sixty feet. On the middle branch of Red river we saw long ranges of bluffs, which, upon examination, proved to be of the same character and composition as those seen on the 3d instant. Soil and subsoil arenaceous.

June 6.—To-day we passed a number of bluffs composed of red clay; I did not observe any gypsum in their composition. As we progressed, the country gradually became more elevated. Here, for the first time since leaving the Witchita mountains, we met with large quantities of drift, composed principally of quartz and mica-schist. On Red river we saw a fine section, fully exposed, showing a horizontal sub-stratum of coarse-grained sandstone, overlaid by drift; the latter forty feet thick.

June 7.—Formation the same as on yesterday.

June 8.—Passed a number of ravines, the sides of most of which were composed of red clay. At about 8 o'clock we came to a small eminence in the prairie, near which I observed an outcrop of grayish-yellow sandstone, presenting a dip of forty degrees to the west. The surface was thickly covered with drift. I saw a number of boulders composed of coarse and fine conglomerate, the largest of which measured fifteen feet in diameter. At 9 o'clock we came to a small creek, with high banks composed of gray calcareous loam, from which latter I obtained a number of shells, characteristic of the loess formation: *Helix plebeium, Succinea elongata,* &c. Soil barren and sandy; subsoil in some places argillaceous, in others arenaceous.

June 9.—Passed to-day a number of small ravines, the sides of which were composed of red clay, overlaid by sandstone and drift. The surface was in many places covered with sand-hills, varying from ten to fifty feet in height. About 8 o'clock we came to an outcrop of finely laminated red ferruginous sandstone, presenting an irregular dip to the northeast of about thirty degrees, (see Section No. 4.) Soil arenaceous; subsoil in many places argillaceous.

June 10.—Formation the same as on yesterday. We frequently found the sandstone exposed and exhibiting evidences of violent disturbance, the strata being variously fractured, and in some places upheaved in such a manner as to stand almost perpendicular. With the exception of the creek bottoms, the soil was sandy and barren; subsoil the same as before.

June 11.—The surface to-day presented nothing but a succession of hills composed of blown sand, varying in height from ten to one hundred feet. No sandstone or drift was anywhere observed.

June 12.—To-day I observed large quantities of drift, of the same composition as before; through it were scattered small boulders, composed of quartz and mica-schist. The surface was in many places covered with loose fragments of carbonate of lime. The particles composing the drift were frequently thickly coated with the same substance. Soil and subsoil arenaceous.

June 13.—Did not leave camp.

June 14.—Drift and limestone the same as before. About 7 o'clock we came to a small ravine, the sides of which exposed a horizontal stratum of coarse-grained sandstone twenty feet thick. From the drift I obtained specimens of agates, chalcedony, and fossil-wood. Soil and subsoil the same as before.

June 15.—The country travelled over to-day was everywhere divided by ridges and ravines; the former sometimes sloping gradually on either side—at others presenting abrupt precipitous terminations. Besides these, a large number of sand-hills, varying in height from twenty to one hundred feet, were observed. The sandstone was frequently exposed. In a few places I found it interstratified with coarse conglomerate; saw a number of small boulders, composed mostly of greenstone, greenstone porphyry, and trachyte. In the bed of the river I found a large mass of black scoriae and several other specimens of volcanic rocks. Drift the same as on yesterday; soil and subsoil arenaceous.

June 16.—The surface was broken, and presented a number of sand-hills. Saw to-day large quantities of drift, which did not differ in composition from that previously noticed. At about eight o'clock we came to a long range of high bluffs, which, as we afterwards ascertained, marked the borders of the "Llano estacado." They were composed of horizontal layers of drift, sandstone, and yellow clay, (see Section No. 7,) all of which seemed to be rapidly yielding to the weather. At the base of the bluffs I observed a few small boulders composed of greenstone porphyry. Soil and subsoil sandy.

June 20.—During the first part of our route we travelled over a hilly and broken region, consisting for the most part of a succession of sand-hills, varying from ten to one hundred feet in height. At the distance of five miles we reached a gradual ascent, which soon led us to the summit of a high and slightly-rolling plain: over its surface were scattered a great many fragments of white carbonate of lime, as well as drift. From the latter I obtained specimens of agate, chalcedony, &c. During the day I had frequent opportunities of observing the formation, which uniformly consisted of drift, interstratified with horizontal layers of red and yellow clay. Sometimes the drift exhibited a calcareous coating, the same as before described.

June 21.—Passed to-day a number of drift-hills, varying in height from twenty to one hundred feet. The surface was very much divided by ravines, with perpendicular sides, composed mostly of red clay, and varying in depth from ten to fifty feet. Near our encampment I observed a horizontal section of yellow loam, coarse conglomerate, and red clay; the last thickly reticulated with gypsum, and overlaid by a terminating stratum of the same, (see Section No. 8.) Soil and subsoil arenaceous.

June 22.—Passed a large number of drift-hills. The country, as on yesterday, was very rough, and much divided by ravines, some of which were fifty feet deep. Their sides were generally composed of red clay, overlaid by drift; in a few instances they exposed seams of gypsum. From the drift I obtained specimens of fossil-wood, agate, jasper, and a few water-worn fossil shells of the genus *Ostrea*. Soil and subsoil the same as before.

June 23.—To-day we observed the gypsum frequently exposed. It did not differ in character from that previously described, and was always found overlying the red clay. Soil and subsoil arenaceous.

APPENDIX D.—GEOLOGY. 191

June 24.—The surface was in many places thickly strewn with loose fragments of white carbonate of lime. About seven o'clock we reached, after a gradual ascent, a high, level, and very fertile plain, from which we obtained an extensive view of the surrounding country, which was very hilly and divided by numerous ravines. The plain at its termination presented a long line of high bluffs, composed of horizontal strata of drift, finely laminated sandstone, white limestone, conglomerate, gypsum, and red and yellow clay, (see Section No. 9.) The red clay was thickly interstratified with thin seams of gypsum. From the drift I obtained specimens of agates, fossil-wood, jasper, and chalcedony. Soil and subsoil the same as before.

June 25.—The country travelled over to-day was very hilly and broken, being much divided by long, narrow ravines, with nearly perpendicular sides, composed of red clay—some of them being over one hundred feet deep. In many places we were surrounded by high bluffs. The drift was found to be unusually abundant—in some places fifty feet thick, and much coarser than before met with. At about eleven o'clock we came in sight of the valley of the Dogtown river. On either side it was bounded by long lines of bluffs, in composition similar to those previously noticed, and varying in height from one hundred to one hundred and fifty feet. From the drift was obtained specimens of chalcedony, agates, silicified wood, and jasper, besides a large number of shells of the same character as those observed on the 22d instant. The beds of the different streams crossed were covered with black ferruginous sand. Soil good, consisting of a rich black mould; subsoil argillaceous.

June 26.—For the first few miles the country was hilly and very much divided by ravines, some of which were two hundred feet in depth. The strata exposed by them were invariably found to consist of horizontal layers of red clay, gypsum, and drift, each occupying the same relative position as shown in Section No. 9.

June 27.—Formation the same as on yesterday. At ten o'clock we came to Dogtown river, the bed of which was composed of yellow sand, intermixed in some places with red clay, and covered with small shining particles of gypsum. I observed in the drift large quantities of red and yellow jasper. Soil fertile; subsoil argillaceous.

June 28.—Saw a large number of drift-hills, varying in height from fifty to one hundred and fifty feet. After travelling a few miles we again came to the borders of the "Llano estacado," which here presented a long line of bluffs six hundred feet high, and composed of horizontal layers of drift and sandstone, interstratified with white limestone. From the base of the bluffs to the river, the country presented a gradual slope of four hundred feet.

Section No. 10 is intended to represent the geological formation from the river level to the summit of the bluffs; the inferior strata, or those between the base of the bluffs and the river, having been ascertained, from numerous observations, to consist of gypsum and red clay. From the drift I obtained specimens of chalcedony, jasper, granite, and obsidian.

July 4.—The formation as observed to-day consisted of red clay, gypsum, and drift: they were all found occupying the same relative positions as before. Soil mostly fertile; subsoil argillaceous.

July 5.—Observed in the prairie a circular outcrop of finely laminated calcareous sandstone about three hundred feet in diameter, and presenting a quaquaversal dip of forty degrees. Over the prairie were strewn a number of small boulders, variously composed of mica-schist, greenstone, and quartz. Red clay, gypsum, and drift, the same as before. Soil and subsoil arenaceous.

July 6.—Observed a number of hills, varying in height from fifty to one hundred feet; in form they resembled truncated cones, and were composed of horizontal layers of sandstone and red clay. General formation the same as before. Soil in some places fertile; subsoil argillaceous.

July 7.—With the exception of the drift, which appears to be rapidly diminishing in thickness, the formation did not differ from that previously observed. The surface was everywhere whitened with beds of gypsum and loose fragments of carbonate of lime. The former varied in thickness from five to fifteen feet; in it were observed large quantities of selenite.

July 8.—The formation was mostly composed of red clay, with a few local deposites of soft carbonate of lime and dark-colored cellular sandstone. Saw no drift or gypsum to-day. Soil fertile; subsoil argillaceous.

July 9.—Again came in sight of the Witchita mountains: the one nearest to us presented the form of a truncated cone, with an irregular basin-shaped depression upon the summit. The formation everywhere consisted of red clay; in a few places it was overlaid by thin seams of gypsum containing selenite. I observed a number of local deposites of white carbonate of lime. Like the gypsum, it was found overlying the red clay. On our route we passed four conical hills from fifty to seventy feet high, and composed of red clay, interstratified with dark-colored porous sandstone. Observed no drift to-day. Soil dark, and fertile; subsoil argillaceous.

July 10.—Formation the same as on yesterday. Soil dark and fertile; subsoil argillaceous.

July 11.—Formation the same as before.

July 12.—To-day we met with no gypsum. At about 9 o'clock we came to Otter creek; its bed is here, as well as elsewhere, composed of finely laminated sandstone, containing small circular spots of a greenish color. In many places this was covered to the depth of a few inches with drift and detritus. Soil fertile; subsoil argillaceous.

July 14.—Renewed the observations of May 23d, 24th, 25th, 26th, and 27th.

July 15.—To-day we passed a number of the Witchita mountains, but observed neither in their composition nor general appearance anything different from what had been previously noticed. Near the base of one of them I observed a nearly horizontal stratum of sandstone, underlaid by red clay. The ground was in several places covered with loose fragments of gypsum, some of which were found to contain slight traces of copper. In one of the creeks I observed a small deposite of black ferruginous sand. Soil black and fertile; subsoil argillaceous.

July 16.—The only difference presented by the mountains seen to-day, from those previously observed, consisted in the greater number and size of the quartz veins: many of them were nearly perpendicular, and extended from near the base of the mountains to their summits; while others, pursuing a more or less serpentine course, frequently intersected each other at right-angles. The largest was highly ferruginous, presented a more or less cellular structure, and was nearly three feet wide. A few feet from the base of one of the mountains I observed a horizontal stratum of coarsely laminated sandstone, of a yellowish color, and including in its composition small angular fragments of granite of the same character as that of the neighboring mountains. To-day I examined several of the head branches of Cache creek. Their beds were thickly strewn with large angular fragments of quartz, greenstone, and porphyry. In each of them I observed large quantities of black ferruginous sand. Soil fertile; subsoil argillaceous.

July 17.—In a number of places the sandstone was exposed; it did not differ in character and composition from that seen the day before. In one place the strata, still preserving their horizontal character, presented abruptly to the side of a mountain. Many of the mountains presented a marked difference in character and composition from any that had been previously observed: instead of displaying a rough and broken exterior, they were more or less rounded, and exhibited a gradual slope to the prairie-level, while the granitic structure almost entirely disappeared, its place being occupied by that of fine porphyry of a reddish color. Scattered over the prairie were observed a great many fragments of granite, greenstone porphyry, and quartz. The beds of the different creeks were in many places covered with black ferruginous sand, as well as large fragments of quartz, porphyry, and hornblende. Soil black and very fertile; subsoil argillaceous.

July 18.—The mountains presented the same appearance and structure as on yesterday. At about 8 o'clock we arrived at Cache creek; its bed was thickly strewn with black ferruginous sand and large fragments of igneous rock. From the water's edge rose abruptly a long line of smooth perpendicular cliffs, varying in height from three to four hundred feet, and having in some places a slight columnar structure, (see Section No. 11.) Upon examination they were found to be composed mostly of fine porphyry of a reddish color, which was traversed by parallel and nearly perpendicular veins of cellular quartz, varying in thickness from two to three feet. Upon its exterior the quartz presented a deep iron-rust color; but when recently fractured, it exhibited various shades of gray and brown, together with small shining particles of sulphuret of iron. Soil fertile, and in some places three feet thick; subsoil argillaceous.

July 19.—I spent the greater part of the day in exploring Cache creek. About one mile below our present encampment I came to the termination of the cliffs. A short distance below this I observed a nearly horizontal stratum of coarsely laminated sandstone, fifty feet thick, and including in its composition fragments of igneous rock of the same character as that composing the cliffs; the intermediate space being occupied by red clay, which, as before, appeared to underlie the sandstone, (see Section No. 11.)

July 20.—Two miles below our camp of last evening I observed a section composed of horizontal layers of gray sandstone, containing in its composition small fragments of igneous rock. Six miles from this we struck a seam of gypsum, varying in thickness from six to twelve inches. Soil fertile; subsoil composed of red and yellow clay.

July 21.—During the day we met with frequent exposures of the sandstone and gypsum. They presented, however, nothing different from what has already been described. Soil and subsoil the same as on yesterday.

July 22.—Formation the same as on yesterday.

July 23.—Did not move from camp.

APPENDIX D.—GEOLOGY.

July 24.—The sandstone appears to be gradually becoming more abundant, while the red clay is less frequently observed than before. Saw no gypsum to-day. Soil fertile, and in some places six feet deep; subsoil composed of yellow clay.

July 25.—Passed a number of small conical hills composed of red clay, overlaid by sandstone. The latter was highly ferruginous, and contained nodular concretions of iron. Soil and subsoil the same as on yesterday.

July 26.—Formation the same as before.

July 27.—At about eight o'clock we came to an extensive outcrop of bluish-gray, non-fossiliferous limestone, which presented in many places a highly crystalline structure. Its relative position was found to be below that of the sandstone. Passed a number of hills, varying in height from one to two hundred feet, and composed of limestone, overlaid by finely laminated sandstone. Soil fertile; subsoil the same as before.

July 28.—To-day the sandstone disappeared almost entirely, its place being occupied by limestone of nearly the same character as that encountered yesterday. Soil and subsoil the same as before.

July 29.—Remained in camp.

July 30.—To-day we again observed the limestone in great abundance. It presented nothing different in character from that previously described. The sandstone and red clay were also in many places largely developed. Soil very fertile; subsoil the same as before. Started from camp at four o'clock in the afternoon. For the first few miles we found the sandstone largely developed; after passing which, we came to an outcrop of limestone of the same character as that previously noticed. It presented itself even with the surface at an angle of thirty degrees, and was over a mile wide. Immediately beyond this we came to a deposite of coarse granite of a reddish color, and variously traversed by veins of quartz. This remarkable formation (as I have been informed) extends about twenty-six miles in an easterly and westerly direction, and is nearly six miles broad. Throughout its entire extent it is said to present the same character, and is everywhere surrounded by aqueous strata. I observed to-day in one of the creeks several boulders, composed of milky quartz; the largest was four feet in diameter. Soil and subsoil the same as before described.

July 31.—Shortly after starting this morning we again struck the limestone formation, which continued to be largely developed during the remainder of the distance to Fort Washita. In it I observed a large number of the characteristic fossils of the cretaceous period.

APPENDIX E.

PALÆONTOLOGY.

DESCRIPTION OF THE SPECIES OF CARBONIFEROUS AND CRETACEOUS FOSSILS COLLECTED: BY B. F. SHUMARD, M. D.

APPENDIX E.

PALÆONTOLOGY.

BY B. F. SHUMARD, M. D.

FOSSILS OF THE CARBONIFEROUS SYSTEM.

CRINOIDEA.

CYATHOCRINUS GRANULIFERUS, Yandell and Shum., mss.
PALÆONTOLOGY, Pl. —, fig. —.

The collection contains a single pentagonal plate of this beautiful encrinite, a perfect specimen of which we found several years since, in the superior carboniferous strata near the summit of Muldrow's Hill, in Kentucky. The costal plate from Arkansas exhibits granules regularly dispersed in rows over the surface, which radiate from the centre to the sides of the pentagon.

It occurs in Washington county, Arkansas, in grayish earthy limestone, associated with *Productus punctatus*, *Terebratula subtilita*, and *Spirifer striatus*.

AGASSIZOCRINUS DACTYLIFORMIS, Troost, mss.
PALÆONTOLOGY, Pl. 1, fig. 7.

Cup conical, composed of three series of pieces; plates massive, smooth, moderately convex; column none; pelvis composed of five pieces, quadrangular, greatest width near the upper edges; second series of pieces five, pentagonal, length and breadth about equal; length of pelvis three lines, greatest width five lines; length and breadth of second series of pieces about three lines.

We regret that the specimens of this crinoid from Arkansas are all imperfect, consisting only of detached portions of the cup. It is a fossil peculiar to the western and southwestern States, and eminently characteristic of the superior members of the carboniferous strata, occurring in some localities very abundantly. The genus is remarkable, from the fact of its being destitute of a column, in which respect it differs from all known carboniferous crinoids. In young individuals, the division of the pelvis into five pieces is well marked; but in adult age they are usually firmly anchylosed, and often all traces of sutures are obliterated. In the centre of the pelvis we observe a small cylindrical tube running nearly its whole length, closed below, but communicating above with the cavity of the cup by a small opening. This structure, probably the nucleus of a column, is only visible when the plates are separated.

It is associated with the preceding species, in the carboniferous beds of Washington county, Arkansas.

PENTREMITES FLOREALIS, Say.

Pentremites florealis, Say, 1820, Jour. Acad. Nat. Sciences, IV, 295,
Pentatrematites florealis, Roemer, 1852, Monog. Blastoid. p. 33, taf. i. fig. 1—4, taf. ii, fig. 8.

This well-known species·is quite common in Washington and Crawford counties, Arkansas. The specimens furnished by my brother are rather more globose than those from localities in Kentucky, Indiana, and Illinois.

PENTREMITES SULCATUS, F. Roemer.

Pentatrematites sulcatus, F. Roemer, 1852, Monog. Blastoid. p. 34, taf. iii, fig. 10, a—c.—*Id*. Lethaea Geognostica, taf. iv, fig. 8, a, b.

We have some doubts as to whether this pentremite is entitled to rank as a distinct species, or whether it should be regarded as merely a variety of *P. florealis*, which varies considerably in different localities. The form under consideration has generally been referred to *P. globosus*, Say, by western geologists, from which, however, it is quite different. Mr. Say's description of *P. globosus* was drawn from a specimen which was brought from Bath, England.

It is associated with the preceding species in Washington county, Arkansas.

BRYOZOA.

ARCHIMEDIPORA ARCHIMEDES, Lesueur.

PALÆONTOLOGY, Pl. 1, fig. 6.

Retepora archimedes, Lesueur, 1842, Amer. Jour. Science, XLIII, 19, fig. 2.

Archimedipora archimedes, D'Orbig. 1849, Prod. de Palæont. I, 102.

A fossil peculiar to the carboniferous strata of the western and southwestern States. The associate fossils are *Pentremites florealis, Productus punctatus, Spirifer striatus,* and *Orthis Michelini.* D'Orbigny, in his Prodrome de Palæontologic, cites this fossil from rocks of the Devonian period, in Kentucky. This is an error; we believe it has not been found lower in the series than the encrinital limestones which repose on the fine-grained micaceous sandstones of the knobs of Kentucky and Tennessee.

It occurs in dark-grayish carboniferous limestone, in Washington county, Arkansas.

BRACHIOPODA.

PRODUCTUS PUNCTATUS, Martin.*

PALÆONTOLOGY, Pl. 1, fig. 5, and Pl. 2, fig. 1.

This *Productus* has a wide geographical, as well as vertical, range in the United States; it is also widely distributed throughout Europe. In this country we find it commencing with the earliest carboniferous deposites, and extending through all the limestones of this system to the coal measures.

Figure 1 of plate 2 represents the ventral valve of a specimen from Washington county, Arkansas; and figure 5 of plate 2, an individual showing the hinge line and the form of the beak.

---

* For synonyms and references, vide Koninck's Monog. du Genre Productus et Chonetes, p. 123.

Productus cora, D'Orbig.

Prod. cora, D'Orbig., 1842, Palæont. Voy. dans l'Amer. Merid., p. 55, pl. 5, fig. 8, 9, 10.

P. tenuistriatus, Verneuil, 1845, Geol. Russ. et Ural., vol. 2, p. 260, pl. 16, fig. 6.

P. cora, Koninck, 1847, Monog. du Genre Prod. et Chonetes, p. 50, pl. iv, a, b, et pl. v, fig. 2, a—d.

The specimens from Arkansas are all imperfect, yet they are plainly referable to this species. It is one of the most characteristic fossils of the carboniferous beds of Kentucky, Indiana, Illinois, and Missouri. It occurs in Washington and Crawford counties, in gray sub-crystalline limestone.

Productus costatus, Sowerby.*

Palæontology, Pl. 1, fig. 2.

This *Productus* occurs with the preceding species, in Washington county, Arkansas, and, like it, has a very extended geographical range in this country and Europe.

Terebratula subtilita, Hall.

Palæontology, Pl. 4, fig. 8.

Terebratula subtilita, Hall, Stansbury's Expedition to Great Salt Lake, 409, pl. xi, fig. 1, a—b, 2, a—c.

This shell is very common in the superior members of the carboniferous formation in Illinois, Indiana, and Kentucky, where it usually is found with *Archimedipora archimedes*, *Pentremites florealis*, and *Productus punctatus*. Its vertical range being rather limited, it constitutes one of our most useful guides in studying the relative position of the various members of the carboniferous strata. This shell is very variable in its characters, so that we are liable to multiply species from its varieties, unless a number of specimens are under examination. Some individuals are very' much inflated; the dorsal valve exhibits a profound

---

* For synonyms and references see Koninck's Monog. du Gen. Prod. et Chonetes, p. 92.

sinus, and the ventral valve a correspondingly elevated ridge. Others are depressed, with scarcely any sinus or bourrelet. The specimens I have seen from Arkansas are considerably mutilated. Occurs in Washington county.

TEREBRATULA MARCYI, Shumard.
PALÆONTOLOGY, Pl. 1, fig. 4, a, b.

Shell small, ovate, elongate, moderately convex, sides and front neatly rounded; dorsal valve regularly convex, rather more gibbous than the opposite valve, greatest height near the beak, no traces of sinus; beak elongated, elevated incurved, no perforation visible in our specimens; ventral valve without median ridge, pointed at summit, cardinal border slightly sinuous. Surface of each valve marked with from thirty-four to thirty-eight simple rounded striæ, which commence at the beak and proceed to the lateral borders and front with division. In general form it resembles *T. serpentina* of Koninck, (Descr. des Animaux fossiles, 29, pl. xix, fig. 8, a—e,) but its smaller size and the lesser number of striæ will serve to distinguish it.

It occurs with *Terebratula subtilita* and *Productus punctatus* in Washington and Crawford counties, Arkansas, in dark-grayish carboniferous limestone. We have found the same species in Floyd county, Indiana.

SPIRIFER, (indet.)
PALÆONTOLOGY, Pl. 1, fig. 3.

In the collection from Washington county are several casts of a spirifer like that which we have figured. They are all too imperfect for description.

FOSSILS OF THE CRETACEOUS PERIOD.

MOLLUSCA.

PECTEN QUADRICOSTATUS, Sowerby.

PALÆONTOLOGY, Pl. 3, fig. 6, and Pl. —, fig —.

Janira quadricostata, D'Orbig. Pal. Franç., III, 644, pl. ccccxlvii, fig. 1—7.

Pecten quadricostatus, F. Roemer, Kreid. Texas, 64, taf. viii, fig. 4, a—c.

Shell sub-ovate, angulated, convexo-concave. Inferior valve convex, with prominent rounded radiating ribs, crossed by five concentric thread-like striæ. Ribs from fifteen to seventeen, of which five are more prominent than the others; smaller ribs disposed in pairs in the intervals between the larger ones. Superior valve slightly concave, with radiating unequal ribs.

As we have not been able to consult Sowerby's description of *Pecten quadricostatus*, we refer our fossil to this species on the authority of Dr. F. Roemer, whose figures and descriptions of specimens from Fredericksburg, Texas, correspond very accurately with those we figure from Fort Washita.

Figure 6 of plate 3 represents the inferior valve of a large individual from Fort Washita, and fig. — of plate — the superior valve of a smaller specimen.

EXOGYRA PONDEROSA, Roemer.

Exogyra ponderosa, F. Roemer, 1849, Texas, 394.
Ostrea ponderosa, D'Orbig., 1850, Prod. de Palæont., II, 256.
Exogyra ponderosa, F. Roemer, Kreid. Texas, 71, taf. ix, fig. 2, a—b.

Shell thick, ovate, sub-cuneiform; large valve gibbous, obtusely carinated, surface marked with imbricating lamellæ; small valve rather thin, sub-concave, surface uneven, concentrically laminated. Occurs rather abundantly at Fort Washita, generally in a fine state of preservation. Roemer cites this species from New Braunfels, Texas.

GRYPHÆA PITCHERI, Morton.

PALÆONTOLOGY, Pl. 6, fig. 5.

Gryphœa Pitcheri, Morton, Synops. Cretaceous Group, 55, pl. xv, fig. 9.
Ostrea vesicularis, D'Orbig. Prod. de Palæont. II, 256, (*pars.*)
Gryphœa Pitcheri, Roemer, Kreid. Texas, 73, taf. ix, fig. 1, a—c.

Shell ovate, thick, gibbous, irregular; inferior valve boat-shaped, inflated, divided into two unequal lobes by a longitudinal furrow, which begins at the umbo and runs the whole length of the shell; umbo large, elongate, incurved and slightly compressed laterally. Superior valve irregular, sub-oval, nearly plane, marked with concentric imbricating lamellæ. Occurs in great numbers in the cretaceous clays at Fort Washita, and more sparingly at Cross-Timbers, Texas. Dr. Morton's specimens were obtained from the plains of Kiamesha, Arkansas, and Dr. F. Roemer found it quite common near New Braunfels, Texas.

EXOGYRA TEXANA, Roemer.

PALÆONTOLOGY, Pl. 5, fig. 1, a—b, and fig. 5.

Exogyra Texana, F. Roemer, Texas, 396.
Ostrea matheroniana, (*pars.*) D'Orbigny, Prod. de Palæont. II, 255.
Exogyra Boussingaultii, Conrad's Geolog. Report of Lynch's Expedition to Dead Sea, 213, pl. i, fig. 9, pl. ii, fig. 10 and 11.
Exogyra Texana, Roemer, Kreid. Texas, 69, taf. x, fig. 1, a—e.

The specimens of this shell in the collection were obtained by Dr. G. G. Shumard, at Camp No. 4, Cross-Timbers, Texas. They vary very much in their characters, scarcely any two examples being alike. In some the shell is quite thin, in others massive; some exhibit prominent rugose ribs, while in others the ribs are but slightly elevated and nodulose. According to Dr. Roemer, this *Exogyra* characterizes the cretaceous deposites near Fredericksburg and New Braunfels, Texas. Mr. Conrad figures a shell from Syria, which he refers to *Exogyra Boussingaultii*, D'Orbig., and which appears to be identical with the species under consideration.

OSTREA SUBOVATA, Shumard.

PALÆONTOLOGY, Pl. 5, fig. 2.

Sub-ovate, trigonal, elongate, massive; inferior valve irregularly convex, inflated, thick, umbo obtusely angulated, somewhat prominent;

ribs four or five, longitudinal, irregular, rounded, nodulose; surface marked with concentric imbricating lamellæ; superior valve rather thin, ovate, nearly plane, slightly convex near the beak, surface with four or five well marked longitudinal undulating sulci.

It occurs at Fort Washita with *Gryphæa Pitcheri* and *Ammonites vespertinus*. It appears to be quite rare, the specimen figured being the only one furnished by the expedition.

INOCERAMUS CONFERTIM-ANNULATUS, Roemer.
PALÆONTOLOGY, Plate 6, fig. 2.

Inoceramus confertim-annulatus, F. Roemer, Texas, 402. Kreidebild. Texas, 59, taf. vii, fig. 4.

Shell ovate, depressed with close concentric undulating ribs; ribs prominent, rounded, regular, intervals about equal to width of ribs.

I refer this fossil to the above species with some hesitation, as all the specimens of the collection are either weather-worn or badly mutilated. Nevertheless, if not identical, ours is a closely allied species. Occurs rather abundantly at Camp No. 4, Cross-Timbers, Texas. Dr. F. Roemer's specimens are from the Guadalupe, near New Braunfels.

TRIGONIA CRENULATA, Lamarck.
PALÆONTOLOGY, Pl. 4, fig. 1.

Trigonia crenulata, Roemer, Kreidebild. Texas, 51, taf. vii, fig. 6.

Shell trigonal, thick, with from fourteen to fifteen oblique crenulated ribs in each valve; anterior side wide, rounded, inflated, posterior side produced, compressed; inferior margin rounded.

From Cross-Timbers, Texas. All the examples in the collection are internal casts. Roemer cites this species from New Braunfels.

ASTARTE WASHITENSIS, Shumard.
PALÆONTOLOGY, Pl. 3, fig. 3.

Shell ovate, trigonal, a little longer than wide, compressed, inequilateral, marked with fine concentric rounded striæ; buccal side shorter than the anal, excavated; basal margin rounded, truncated posteriorly, beaks slightly prominent, excavated.

The only specimen of this species collected by the expedition is rather too imperfect to permit us to make a satisfactory description. It was found in the cretaceous strata at Camp No. 4, Cross-Timbers, Texas.

Cardium multistriatum, Shumard.
PALÆONTOLOGY, Pl. 4, fig. 2.

Shell sub-rotund, inflated, length and breadth nearly equal, truncated posteriorly, basal and anterior margins rounded; surface of posterior sub-margin with from fourteen to fifteen regular radiating striæ; remainder of surface marked with fine, equal, rounded, close, concentric striæ. Beaks rather prominent.

This is a neat, pretty species; and it is to be regretted that the specimens collected werè not in a better state of preservation. It was found at encampment No. 4, Cross-Timbers, Texas, where it is rather uncommon.

Panopæa texana, Shumard.
PALÆONTOLOGY, Pl. 6, fig. 1.

Shell oval, elongate, inflated anteriorly, compressed behind, beaks moderately prominent, basal edge rounded, buccal extremity wide, rounded; surface marked with irregular concentric slightly elevated ribs. Length about 2 5-10 inches, breadth 1 4-10 inch, thickness 1 1-0 inch.

The only specimen of this species brought home by the expedition is an imperfect cast.

Locality, encampment No. 4, Cross-Timbers, Texas.

Terebratula choctawensis, Shumard.
PALÆONTOLOGY, Pl. 2, fig. a, b.

Shell sub-globose, inflated, sub-pentagonal, front slightly truncated, surface minutely punctate, the puncti only visible when examined through a strong lens; dorsal valve most inflated; beak obtuse, recurved, pierced by an oval aperture; area distinct, forming a well defined obtuse angle; ventral valve moderately convex, sub-orbicular. Length 9 lines, width 8 lines, thickness $6\frac{1}{2}$ lines.

It resembles *Terebratula wacoensis*, (F. Roemer, Kreidebild. Texas, 81, taf. vi, fig. 2, a–c,) but differs in the character of the surface, which in *T. Choctawensis* is thickly studded over with minute puncta. It is also a smaller species; the area is not so wide comparatively, and the front is not so broadly truncate.

This beautiful Terebratula was obtained from the cretaceous deposites near Fort Washita, where it is quite rare, a single specimen only having been found.

GLOBICONCHA (TYLOSTOMA) TUMIDA, Shumard.
PALÆONTOLOGY, Pl. 5, fig. 3.

Shell ovate-globose, spire pyramidal, volutions about six, whorls moderately convex; width of body whorl equal to about one half the length of the shell. Length 1 7-10 inch, width 1 3-10 inch.

All the specimens we have seen are badly preserved internal casts. Occurs at Cross-Timbers, Texas, in cretaceous limestone.

GLOBICONCHA (?) ELEVATA, Shumard.
PALÆONTOLOGY, Pl. 4, fig. 4.

Shell ovate, spire produced, whorls six regularly convex, body whorl shorter than spire. Length 1 5-10 inch, breadth 1 inch.

This is likewise an internal cast. It occurs with the preceding species.

EULIMA (?) SUBFUSIFORMIS, Shumard.
PALÆONTOLOGY, Pl. 4, fig. 3.

Shell subfusiform, elongate smooth, spire produced, regularly conical; whorls about six, broad, very sllghtly convex; suture rather shallow, linear, aperture simple, sub-ovate; body whorl obtusely angulated. Length 2 8-10 inches, width 1 1-10 inch.

The collection contains only a single specimen of the cast of this species, and that badly weather-worn. It was found at Camp No. 4, Cross-Timbers, Texas.

AMMONITES VESPERTINUS, Morton.

Ammonites vespertinus, Morton, Synopsis Cretaceous Group U. S., 40, pl. xvii, fig. 1. *Id.*, D'Orbigny, Prodrome de Palæont. II, 212.

Shell large, volutions about three; vertical section sub-quadrangular; ribs prominent, each garnished with three nodules, dorsal one most prominent; dorsal margin furnished with a prominent rounded carina.

This is the largest species of Ammonite that has hitherto been found in the United States. In the cretaceous strata near Fort Washita, specimens were found to measure nearly three feet in diameter, and estimated to weigh upwards of two hundred pounds. It is quite common. The fragment described by Dr. Morton was obtained from the plains of Kiamesha, Arkansas.

AMMONITES MARCIANA, Shumard.
PALÆONTOLOGY, Pl. 4, fig. 5.

Shell compressed, not carinated, with about twelve simple, prominent rounded ribs, which cross the dorsum and sides of the last volution obliquely, without interruption; dorsum convex, whorls compressed; surface smooth in the intervals between the ribs; aperture longitudinal, sub-oval.

Length of last whorl 11 lines, width of do. $4\frac{1}{2}$ lines; width of umbilicus 3 lines.

The specimen figured is a cast, and the character of the lobes of the chambers cannot be made out.

Occurs in the cretaceous strata of Cross-Timbers, Texas.

AMMONITES ACUTO-CARINATUS, Shumard.
PALÆONTOLOGY, Pl. 1, fig. 3.

Shell much compressed, sharply carinated, ornamented with from 30 to 34 transverse ribs; ribs simple, distinctly elevated, flexuous commencing narrow at the umbilicus and widening to within a short distance of the dorsal border, where they are again somewhat contracted; dorsal carina prominent, sharp smooth, marked on each side by a shallow depression; aperture elongate-cordate, lateral septa trilobate.

Diameter 2 4-10 inches; thickness of last whorl near aperture 5-10 inch.

AMMONITES—(*undetermined.*)

PALÆONTOLOGY, Pl. —, fig. —.

Several specimens of a small variety of ammonite, such as is represented in the figure, were found with the two last-described species, but they are too imperfect for satisfactory description.

ECHINODERMATA.

HEMIASTER ELEGANS, Shumard.

PALÆONTOLOGY, Pl. 2, fig. 4, a, b, c.

Shell ovate orbicular, moderately convex at summit, broadly emarginate anteriorly; anal extremity truncated almost vertically, very slightly excavated; ambulacra sub-petalloid, broad, situated in shallow depressions, antero-lateral areas widely divergent, extending to the margin of the test, postero-lateral areas much less divergent and short, peripetalous fasciole indistinct; mouth transverse reniform, not far from the anterior border, post oral tuberculated space lanceolate; anus oval longitudinal, sub-anal fasciole scarcely visible; surface of test covered with small spinigerous tubercles, with minute granulæ in the interspaces. The dimensions of the largest specimen that I have been permitted to examine are as follows: length, 2 7-10 inches; greatest width, 2 5-10 inches; height, 1 5-10 inch.

This exceedingly elegant species occurs in great numbers in the cretaceous strata at Fort Washita.

HOLASTER SIMPLEX, Shumard.

PALÆONTOLOGY, Pl. 3, fig. 2.

Shell ovate, sub-cordate, gibbous, regularly rounded superiorly, most prominent near apex, which is sub-central, declining at first gently towards the mouth, then abruptly, truncated posteriorly, with a thread-like carina leading from the apex to the anus; oral sinus shallow, rounded;

Occurs rather abundantly with the preceding species at Cross-Timbers, Texas.

ambulacra flexuous, extending to the base, increasing gradually in width to the inferior margin; antero-lateral are aswidely divergent; postero-laterals separated by a moderate interval; mouth transverse, oval; anus oval, longitudinal sub-anal fasciole indistinct; surface of test sparingly studded with spinigerous tubercles, with numerous microscopic granules in the interspaces. It approaches *Holaster (Ananchytes) fimbriatus*, Morton, (Silliman's Journal, XVIII, 245, pl. 3, fig. 9.) Our specimens, however, differ from the figures given by Dr. Morton in being less orbicular in the oral sinus, which is not so profound, and in the anal border, which is more widely truncated

Occurs with *Hemiaster elegans* at Fort Washita.

HOLECTYPUS PLANATUS, Roemer.

Holectypus planatus, F. Roemer, Texas, 393. *Ibid*, Kreidebild. Texas, 84, taf. x, fig. 2, a—g.

In the collection from Cross-Timbers, Texas, we find several mutilated specimens of *Holectypus*, which we refer without doubt to the above species. Dr. Roemer's examples were obtained from the vicinity of Fredericksburg, Texas.

APPENDIX F.

ZOOLOGY.

MAMMALS, BY R. B. MARCY, CAPT. U. S. A.

REPTILES, BY S. F. BAIRD AND C. GIRARD.

FISHES, BY S. F. BAIRD AND C. GIRARD.

SHELLS, BY C. B. ADAMS AND G. C. SHUMARD, M. D.

ORTHOPTEROUS INSECTS, BY C. GIRARD.

ARACHNIDANS, BY C. GIRARD.

MYRIAPODS, BY C. GIRARD.

APPENDIX F.

ZOOLOGY.

MAMMALS.

BY CAPTAIN R. B. MARCY.

URSUS AMERICANUS, Pall. Black bear. Throughout the valley.

PROCYON LOTOR, L. Raccoon. Throughout the valley.

MEPHITIS MESOLEUCA, (?) Licht. Texan skunk. Throughout the valley.

LUTRA CANADENSIS, Sabine. Otter. Throughout the valley.

BASSARIS ASTUTA, Licht. Civet cat. Cross-Timbers.

CANIS OCCIDENTALIS, Rich. Gray wolf. Above Shreveport.

CANIS LATRANS. Prairie-wolf. Above Cross-Timbers.

CANIS ———. Large Lobos wolf. Above Cross-Timbers.

VULPES FULVUS. Red fox. Red river valley.

LYNX RUFUS. Wild cat. Red river valley.

FELIS CONCOLOR, L. Panther. Red river valley.

SCIURUS MAGNICAUDATUS, (?) Say. Fox-squirrel. Red river valley.

TAMIAS QUADRIVITTATUS, Say. Striped squirrel. Above Cross-Timbers.

PTEROMYS VOLUCELLA, Gm. Flying-squirrel. Red river valley.

CASTOR FIBER, L. Beaver. Above Cross-Timbers.

LEPUS SYLVATICUS, Bach. Rabbit. Red river valley.

LEPUS CALLOTIS, (?) Wagl. Jackass rabbit. Above Cross-Timbers.

APPENDIX F.—MAMMALS.

LEPUS ARTEMISIA (?) Small prairie rabbit. Above Cross-Timbers.

SPERMOPHILUS LUDOVICIANUS, Ord. Prairie-dog. Above Cross-Timbers.

DIDELPHYS VIRGINIANA, Shaw. Opossum. Red river valley.

CERVUS VIRGINIANA, Penn. Deer. Red river valley.

CERVUS CANADENSIS. Elk; only about Witchita mountains.

ANTILOCAPRA AMERICANA, Ord. Antelope. Above Cross-Timbers.

BOS AMERICANUS, L. Above Cache creek.

REPTILES.

BY S. F. BAIRD AND C. GIRARD.

SERPENTS.

The serpents collected by Captains Marcy and McClellan belong to ten species, distributed into eight genera. Several of these species had previously been received from other sections of the country: three, however, were first collected during the expedition. All are here figured for the first time, except *Ophibolus Sayi*, of which a hitherto undescribed variety is represented.

I. CROTALUS, Linn.

This genus is characterized by its erectile poison fangs, and by having the upper surface of the head covered with small plates resembling the scales on the body, and with only a few larger ones in front. There is a deep pit between the eyes and the nostrils. The plates under the tail are undivided, and the tail is terminated by a rattle. Scales carinated.

1. CROTALUS CONFLUENTUS, Say.
ZOOLOGY, Pl. 1.

SPEC. CHAR.—Head subtriangular. Plates on top of head squamiform, irregular, angulated, and imbricated; scales between superciliaries small, numerous, uniform. Four rows of scales between the suborbital series (which only extends to the centre of the orbit) and the labials. Labials 15 or 18, nearly uniform. Dorsal series 27–29. Dorsal blotches quadrate, concave before and behind; intervals greater behind. Spots transversely quadrate posteriorly, ultimately becoming 10 or 12 half rings. Two transverse lines on superciliaries, enclosing about one-third. Stripe from superciliary to angle of jaws, crosses angle of the mouth on the second row above labial. Rostral margined with lighter.

Syn.—*Crotalus confluentus,* Say, in Long's Exped. Rocky Mts. II, 1823, 48. B. & G. Cat. N. Amer. Rept. I, 1853, 8.

C. Lecontei, Hallow. Proc. Acad. Nat. Sc. Philad. VI, 1851, 180.

DESCRIPTION.—This species bears a considerable resemblance to *C. atrox,* but the body is more slender and compact. Scales on the top of the head anterior to the superciliaries nearly uniform in size. Line of scales across from one nostril to the other consists of six, not four as in *C. atrox.* Superciliaries more prominent. Labial series much smaller. Upper anterior orbitals much smaller, as also is the anterior nasal. Scales on the top of the head less carinated. Scales between superciliaries smaller and more numerous, five or six in number instead of four. Two lateral rows of scales smooth; first, second, and third gradually increasing in size. Scales more linear than in *C. atrox.*

General color yellowish brown, with a series of subquadrate dark blotches, with the corners rounded and the anterior and posterior sides frequently concave, the exterior convex. These blotches are ten or eleven scales wide and four or five long, lighter in the centre, and margined for one-third of a scale with light yellowish. The intervals along the back light brown, darker than the margins of the blotches. Anteriorly the interval between the dark spots is but a single scale; posteriorly it is more, becoming sometimes two scales, where also the spots are more rhomboidal or lozenge-shaped; nearer the tail, however, they become transversely quadrate. The fundamental theory of coloration might be likened to that of *Crotalus adamanteus,* viz: of forty or fifty light lines decussating each other from opposite sides; but the angles of decussation, instead of being acute, are obtuse, and truncated or rounded off throughout. Along the third, fourth, and fifth lateral rows of scales is a series of indistinct brown blotches covering a space of about four scales, and falling opposite to the dorsal blotches: between these blotches, and opposite to the intervals of the dorsal blotches, are others less distinct. Along the fifth, sixth, seventh, and eighth rows is a second series of obsolete blotches, each covering a space of about four scales, and just opposite the intervals between the dorsal spots. The dorsal and lower series are separated by an interval of three scales, this interval light brown. Beneath, the color is dull yellowish, and ten or twelve darker half rings are visible on the tail.

In point of coloration the principal features, as compared with *C. atrox,* lie in the disposition of the dorsal blotches in subquadrate spots instead of subrhomboids; the intervals thus forming bands across the back perpendicular to the longitudinal axis. This tendency to assume the subquadrangular pattern has broken up the chain-work into isolated portions, as in *Ophibolus eximius* or *Crotalophorus tergeminus*. The intervals of the dorsal blotches are wide and darker in the middle, while in *C. atrox* they are narrow, not linear, and unicolor. The sides of the head present the usual light stripe from the posterior extremity of the superciliary; it passes, however, to the angle of the jaw on the neck, along the second row of scales above the labials. A second stripe passes in front of the eye to the labials, widening there. A small light vertical bar is seen below the pit, and another on the outer edge of the rostral. On the superciliaries are two light transverse lines enclosing a space nearly one-third of the whole surface. In *C. atrox* there is a single median line. Sometimes, as in *C. atrox*, the single blotches on the nape are replaced by two elongated ones parallel to each other.

Dorsal row of scales, 29; abdominal scutellæ, 180; subcaudal ones, 27. Total length, 34 inches; length of tail, 4 inches.

A specimen was collected the 5th of June in the Witchita mountains. Another specimen of the same species was brought home from the Cross-Timbers, Arkansas, by Dr. S. W. Woodhouse, and described by Dr. Hallowell as new, under the name of *Crotalus Lecontei*, on the ground that the anterior vertebral spots are not confluent. This we do not consider as a sufficiently distinctive character, although we have never seen a specimen with decidedly confluent markings. The notes of Dr. Leconte, quoted by Dr. Hallowell, hardly apply to the present species.

The species was first discovered by Say on Major Long's expedition to the Rocky mountains, and has not since been seen until procured first by Dr. Woodhouse, and then by Captain Marcy and the Mexican boundary commission. It was found by the latter party in Western Texas, where, however, it is rare.

Plate I represents *Crotalus confluentus* of natural size.

II. EUTÆNIA, B. & G.

This genus is composed of numerous species, some of them quite common, and known under the names of Riband, Striped, and Garter snakes: inoffensive, like most of the North American snakes. They may be recognised by three light stripes on a darker ground, the intervals between these stripes provided with alternating or tessellated blackish spots. The scales have a ridge or small keel along their middle, and are arranged in 19 or 21 longitudinal rows. The postabdominal or anal scutella is entire, like the others. There is one anterior orbital plate and three posterior. The body is either moderately stout or else slender, according to the species. Of the two described in this article, one belongs to the division with a slender body and 19 dorsal rows of scales, and the other to the second division, with a stouter body and 21 dorsal rows of scales.

2. EUTÆNIA PROXIMA, B. & G.

ZOOLOGY, Pl. II.

SPEC. CHAR.—Body stoutest of the division. Black above; three longitudinal stripes, the dorsal ochraceous yellow or brown, lateral greenish white or yellow. Total length about three and a half times that of the tail.

SYN.—*Coluber proximus*, Say, in Long's Exped. to Rock. Mts. I, 1823, 187.—Harl., Journ. Acad. Nat. Sc. Philad. V, 1827, 353.
Eutainia proxima, B. & G. Cat. N. Amer. Rept. I, 1853, 25.

DESC.—Deep brown, almost black, above and on the sides; beneath greenish white. Dorsal stripe on one and two half rows of scales, ochraceous yellow, lateral stripe on the 3d and 4th rows of scales, greenish yellow or white, markedly different in tint from the dorsal. Sides of abdominal scutellæ, and 1st and 2d dorsal series, of the same color as the back. On stretching the skin, numerous short white lines are visible. Occipital plates with two small approximated spots on the line of junction. Orbitals whitish. The greenish white of the abdomen becomes more yellow anteriorly.

Head more like that of *E. saurita* than of *E. Faireyi*, while the body is stouter than in either. The subcaudal scales are less numerous than in the other two allied species. Resembling *E. Faireyi* in color, it is always distinguishable by the stouter body, fewer caudal scales, and dissimilarity of color in the longitudinal stripes.

Dorsal rows of scales 19, all keeled; abdominal scutellæ 170; subcaudal ones 100. Total length 33 inches; length of tail 9 inches. Found at Camp No. 7.

The species is represented in natural size on Plate II.

3. EUTÆNIA MARCIANA, B. & G.
ZOOLOGY, Pl. III.

SPEC. CHAR.—Prominent color light brown; a vertebral paler line and one lateral on each side, more or less indistinct. Three series of square black spots on each side, of about 56-60 in each series, from occiput to anus. Sides of head black, with a crescentic patch of yellowish posterior to the labial plates. Three and sometimes four black vittæ radiating from the eye across the jaws. A double white spot with a black margin on the suture of occipital plates.

SYN.—*Eutainia marciana*, B. and G. Cat. N. Amer. Rept. I, 1853, 36.

DESC.—The markings about the head are generally very constant and distinct. Viewed laterally, we see first the large dark-brown patch at the back part of the head, extending as far back as the posterior extremity of the jawbones. In the anterior part of this patch is seen the crescentic patch (concave before) of yellowish white, with a more or less narrow dark-brown margin anteriorly. The next black band starts from the posterior edge of the superciliaries, and passes obliquely downwards and backwards along the posterior edge of the 6th upper labial. Similar black margins are seen on the posterior edges of the 5th and 4th labials, the intervening spaces being yellowish white, particularly on the 5th upper labial. Occasionally the posterior margins of the 7th and 3d labials have the black line as well as those mentioned, which frequently extend across to the posterior margins of the corresponding lower labials. The white spot on the anterior portion of the occipital suture is always margined with black.

The six series of black spots are arranged so as to alternate with each other. The lower or third series on each side is below the indistinct lateral stripe. The posterior edge of each abdominal scutella shows a black margined spot on each side. The dorsal line is generally a single scale in width, occasionally including portions of the lateral, and itself sometimes encroached upon by the black spots. Each spot is about a scale or a scale and a half long, and about three scales broad. The number in the dorsal series from the head to the anus varies from 56 to 60. Posterior edges of scales very slightly emarginate, if at all. All are decidedly keeled.

Dorsal scales disposed in 21 rows; abdominal scutellæ, 152; subcaudal, 75. Total length 34 inches; length of tail, 8 inches.

Collected between Camp 5 and Red river, on the open prairie.

This species is very widely distributed in the south and west. Red river forms its limit on the north, and the Gulf of Mexico on the east;

but it extends to the Rocky mountains on the west, and far into Mexico on the south. Its centre of distribution appears to be on the lower Rio Grande.

Plate III represents this species in natural size.

III. HETERODON, Pal. de B.

This genus is eminently characterized by the peculiarity of its snout, which is terminated by a triangular plate recurved upwards; hence the popular appellation of hog-nose snake. Though perfectly harmless, they exhibit a threatening appearance, when approached, in the flattening of their head and violent hissings; hence the names of blowing-viper, spreading adder, &c. Their body is short, stout, and the tail also short. The head is broad and short. The dorsal scales are carinated, and arranged in 23–27 rows. The preanal or postabdominal scutella is bifid; a chain of small plates beneath the eye, completed above by the superciliaries. There is a supplementary plate on the top of the head, behind the prominent rostral, either in contact with the frontals, or separated by smaller plates. The colors are light, with dorsal and lateral darker blotches, or else brown, with dorsal transverse light bars; sometimes entirely black.

One species only was collected on the Red River exploration. Six species are known to exist in the United States.

4. HETERODON NASICUS, B. & G.

ZOOLOGY, Pl. IV.

SPEC. CHAR.—Vertical plate broader than long. Rostral excessively broad and high. Azygos plate surrounded behind and on the sides by many small plates (12–15.) A second loral. Labials short and excessively high. Dorsal rows of scales 23, exterior alone smooth. A dorsal series of about 50 blotches, with four or five other series on each side. Body beneath, black. A narrow white line across the middle of the superciliaries; a second behind the rostral. A broad dark patch from the eye to the angle of the mouth, crossing the two postlabials.

SYN.—*Heterodon nasicus*, B. & G. Reptiles in Stansbury's Expl. Valley of Great Salt Lake, 1852, 352.—B. & G. Cat. N. Amer. Rept. I, 1853, 61.

DESC.—Vertical plate very broad, subhexagonal. Occipitals short. Rostral very broad, high, more so than in the other species, outline rounded. The interval between the opposite frontals, the rostral, and the vertical occupied by a number of small plates, from 10 to 12, or more, arranged without any symmetry, on each side and behind the small azygos. The base of the rostral between the opposite prenasals is generally margined by these small plates, which sometimes, too, are seen between the vertical and the anterior portion of the superciliaries. This crowding of plates causes the anterior part of the forehead to be broader than in *H. simus*. Eye small, its centre rather posterior to the middle of the imaginary line connecting the tip of rostral with the lower angle of the postlabial, which line scarcely crosses the eyeball. Orbital plates, 10–13 in number. Loral triangular, rather longer than high, separated from the frontal by a small plate. Nasals rather short, occasionally with the lower part of the nostril bounded by a small plate. Labials 8 or 9 above, all of them higher than long; indeed, their vertical extension is much greater than in any other species: the 6th highest; centre of eye over the junction of the 5th and 6th.

Dorsal rows of scales 23, outer row smooth, rest all distinctly carinated, the keels extending to the ends of the scales; those just behind the occipital plates truncated, with obsolete carinæ. Scales on the hind part of the body rather broader and shorter than anteriorly; the inequality scarcely evident in large specimens.

Ground-color light brown or yellowish gray, with about 50 dorsal blotches from head to tip of tail; the 39th opposite the anus. These blotches are quite small, rather longer transversely, subquadrate, or rounded, indistinctly margined with black, (obsoletely on the outside;) they cover 7 to 9 scales across, are 2 to 2½ long, and separated by interspaces of 1½ scales, which are pretty constant throughout, though rather narrower on the tail. On each side of the dorsal row may be made out, under favorable circumstances, four alternating rows of blotches; the first on the contiguous edges of the scales of the first and second exterior dorsal rows; the second on the scales of the 3d row, and the adjacent edges of those in the 2d and 4th ; the third on the scales of the 4th, 5th, and 6th, and the adjacent edges of the 3d and 7th; and the fourth on the scales of the 6th, 7th, and 8th rows, and the adjacent edges of those of the 5th. This last is opposite the intervals of the dorsal series; the rest alternate with it. The central inferior surface of the abdominal scutellæ is black, sharply variegated with quadrate spots of yellowish white; the portion of the scutellæ entering into the side of the body is yellowish white, with that part opposite the dorsal intervals dark brown, thus, in fact, constituting a fifth lateral series of blotches, alternating with the lowest already mentioned. The throat and chin are unspotted. The head is light brown, with a narrow whitish line finely margined before and behind with black, which crosses in front of the centre of the vertical, and through the middle of the superciliaries: a second similar but more indistinct line runs parallel to this just behind the rostral, and extending down in front of the eye. A third equally indistinct and similar line crosses the posterior angle of the vertical, and runs back on the side of the neck, behind the labials and temporal shields. There is a broad brown patch from the back part of the eye to the angle of the mouth, across the penultimate and last labial. The coloration is thus very different from that of *H. Simus*, where there is a distinct narrow black band across the forehead scarcely involving the vertical, and passing through the eye to the angle of the mouth across the last labial. Behind this a much broader yellowish band, continued without interruption into the neck behind the angle of the mouth. In *H. nasicus* the most conspicuous feature is a narrow white band, much narrower than the darker patch before and behind it. The dark patch, to the angle of the mouth, is much broader, continuous as it were, with the broad bar between the middle and anterior light lines, which corresponds with the narrow black line of *H. Simus*. The other distinguishing features are evident. The three dark patches behind the head are much as in *H. simus*.

In large specimens from Sonora and the Copper Mines of the Gila, (Fort Webster), the ground-color is yellowish gray, each scale minutely punctate with brown. The blotches are all obsolete, only one dorsal and two lateral on each side being defined by darker shades. The blotches on the sides of the abdomen are wanting, but the black in the middle is strongly marked. The other characters, however, are preserved, except that the exterior row of dorsal scales is more or less carinated.

Specimens of this species vary in the number of small postrostral plates. In some there are only three or four, in others a larger number. Sometimes, instead of a single series of median dorsal spots, there are two, in close contact, and more or less confluent. The narrow light line across the middle of the superciliaries and the high labials are always highly characteristic.

The specimen figured of natural size on plate IV, is much smaller than those alluded to from Sonora and the Copper Mines, and upon which the foregoing description has been based.

IV. PITUOPHIS, Holbr.

This genus, closely allied to *Heterodon*, is characterized by a prominent snout, the rostral plate elevated and convex, without, however, being recurved. There are two pairs of postfrontal plates instead of one, and occasionally also two verticals; three or four postorbitals; generally two, sometimes only one anteriorly. The scales are carinated along the back, smooth on the sides, and constituting from 29–35 dorsal rows. The preanal or postabdominal scutellæ is entire.

The ground-colors are either whitish or reddish yellow, with a triple series of patches, those of the medial series the largest, and several series of smaller blotches on the sides. Abdomen unicolor or spotted, with an outer row of blotches. Head of same color as the body, maculated with black spots. A narrow band of black across the upper surface between the eyes, and a postocular vitta on each side, extending obliquely from the eye down to the angle of the mouth. A black vertical patch is often seen beneath the eye.

The names of Bull, Pine, and Pilot snake are commonly given to different species of this genus, which are all of great size, including in fact some of the largest serpents of North America. Some of the species utter a hissing or blowing sound.

5. Pituophis McClellanii, B. & G.

Zoology, Pl. V.

Spec. char.—Head subelliptical. Rostral plate very narrow. Anteorbitals 2; postorbitals 4. Dorsal rows 33–35; the 7 outer rows smooth. Tail forming 1·9 or 1·10 of total length. Postocular vitta brown, and rather broad. Suborbital black patch conspicuous; commissure of labials black. Color of body reddish yellow, with a series of 53 blotches from head to origin of tail. Blotches of adjoining series, on either side, confluent across the light spaces between medial blotches. Flanks covered with small blotches, forming 3 or 4 indistinct series. Twelve transverse jet-black bars across the tail. Abdomen yellowish, thickly maculated with black patches.

Syn.—*Pituophis McClellanii*, B. & G. Cat. N. Amer. Rept. I, 1853, 68.—Pilot-snake.

DESC.—Head proportionally large, ovoid, distinct from the body. Snout pointed. Occipital plates small. Vertical broad, subpentagonal, slightly concave on the sides. Superciliaries large. Internal postfrontals rather narrow, elongated; external postfrontals quadrilateral, a little broader forwards. Prefrontals irregularly quadrangular. Rostral very narrow, extending halfway between the prefrontals, convex and raised above the surface of the snout. Nostrils in the middle line between the nasals, the posterior of which is a little the smaller. Loral trapezoidal, proportionally large. Inferior anteorbital very small, resting upon the fourth upper labial. Postorbitals varying in comparative size. Temporal shields small, resembling scales. Upper labials 8; 6th and 7th the larger. Lower labials 12; 6th and 7th largest. Posterior mental shields very small, extending to opposite the junction of the 7th and 8th lower labials. Scales proportionally small, in 33–35 rows, the 7 outer ones perfectly smooth and somewhat larger than the remaining rows.

Ground-color yellowish brown, with three series of dorsal black blotches, 53 in number, from the head to opposite the anus, with 12 on the tail, in the shape of transverse bars. Those of the medial series the larger, and covering 8 or 9 rows of scales. On the anterior part of the body they are subcircular, embracing longitudinally four scales; posteriorly they become shorter by one scale. The light spaces between are a little narrower than the blotches themselves for the twelve anterior blotches, and wider than the blotches for the remaining length of the body. The blotches of the adjoining series alternate with those of the medial series, being opposite to the light intermediate spaces, across which the blotches of either sides are generally united by a transverse narrow band. The flanks are densely covered with small and irregular blotches, forming three indistinct series, confluent in vertical bars towards the origin of the tail. Inferior surface of the head yellowish, unicolor Abdomen dull yellow, with crowded brownish black blotches in series on the extremity of the scutellæ.

Two specimens of this species were caught the 28th of June. The largest is figured, of natural size, on plate V. It is $38\frac{1}{4}$ inches in length; the tail measuring nearly 5 inches. Abdominal scutellæ 231; subcaudal ones, **52**.

V. SCOTOPHIS, B. & G.

The scales in this genus are very slightly carinated on the back, and perfectly smooth on the sides. Preanal scutella bifid. One large anteorbital plate and two postorbitals. The colors are brown or black, in quadrate blotches on the back and on the sides, separated by lighter intervals; beneath usually coarsely blotched with darker. In one species there are dark stripes on a light ground.

6. SCOTOPHIS LAETUS, B. & G.
ZOOLOGY, Pl. VI.

SPEC. CHAR.—Similar to *S. confinis*, but postfrontals larger. Vertical plate longer than broad. Dorsal rows 29. Abdominal scutellæ 227. Subcaudals 72. Blotches fewer than in *S. confinis*.

SYN.—*Scotophis laetus*, B. & G. Cat. N. Amer. Rept. I, 1853, 78.

DESC.—This species bears a close resemblance to *S. confinis*, and its characters may be best given by comparison with the latter. It differs, therefore, in the greater number of dorsal rows, 29 instead of 25. The whole body and head are much stouter. Exterior eight rows smooth, rest slightly carinated. The vertical is broad before, rather acute behind. A probably monstrous feature is seen in the union of the two postfrontals, except for a short distance before, and in the loral and postnasal coalescing into one trapezoidal plate. Blotches less numerous. A broad vitta across the back part of the postfrontals, passing backwards and downwards through the eye, and terminating acutely on the posterior upper labial. A blotch across the back part of the vertical, and extending through the occipitals on each side to the nape. The spots are larger, longitudinal throughout, with occasional exceptions.

Its affinities to *S. vulpinus* are close. The vertical, however, is narrow, the eyes much larger, dorsal rows 29 instead of 25. The blotches on the back are longitudinal, and fewer in number. For a complete description of this species it will be necessary to procure larger specimens.

The specimen figured on Plate VI is of natural size. The only one caught of this species is 18 inches long. Length of tail $3\frac{1}{4}$ inches.

VI. OPHIBOLUS, B. & G.

The body is rather thick, and the tail short. The scales smooth and lustrous, and disposed in 21 or 23 rows, which scarcely overlap. The preanal scutella is entire. A small anteorbital plate and two postorbitals. Eyes very small.

The ground-colors are black, brown, or red, crossed by lighter intervals, generally bordered by black.

Seven species, besides the two here described, have hitherto been found in North America.

7. Ophibolus Sayi, B. & G.
Zoology Pl. VII.

Spec. char.—Black, each scale above with a large circular or subcircular white or yellow spot in the centre. Sometimes only transverse lines of these spots across the back.

Syn.—*Herpetodryas getulus*, Schl. Ess. Phys. Serp. Part. descr. II, 1837, 198.

Coronella Sayi, Holbr. (non Schl.) N. Amer. Herp. III, 1842, 99. Pl. xxii.

Coluber Sayi, Dekay, New York Fauna, Rept. 1842, 41.

Ophibolus Sayi, B. & G., Cat. N. Amer. Rept. I, 1853, 84.

Desc.—Body, as in most of the other species of the same genus, very tense and rigid, with difficulty capable of being extended after immersion in alcohol. Vertical plate triangular, wider than long; outer edge slightly convex, an angle being faintly indicated at the junction of the superciliaries and occipitals; shorter than the occipitals, which are short, longer than broad. Postfrontals large, broad; prefrontals smaller. Rostral small, not projecting, slightly wedged between prefrontals. Eye very small, orbit about as high as the labial below it; centre of the eye a little anterior to the middle of the commissure, over the junction of the 3d and 4th labials. One anteorbital, vertically quadrate; loral half its height, square. Upper labials 7, increasing to the penultimate. Lower labials 9; 4th and 5th largest.

Scales nearly as high as long, hexagonal, truncated at each end. Dorsal rows 21, exterior rather larger, and diminishing almost imperceptibly to the back, although all the scales in a single oblique row are of very nearly the same shape and size.

The scales on the back and sides are lustrous black, each one with a central elliptical or subcircular spot of ivory-white, which on the sides occupy nearly the whole of the scale, but are smaller towards the back, where they involve one half to one third of the length. Beneath yellowish white, with broad distinct blotches of black, more numerous posteriorly. Skin between the scales brown. The plates on the top and sides of the head have each a yellowish blotch; the labials are yellow, with black at their junction.

Other specimens agree except in having bright yellow instead of white as described; the spots, too, are rather smaller, and manifest a slight tendency to aggregation on adjacent scales, so as to form transverse bands. This is seen more decidedly where the back is crossed by about 70 short dotted yellow lines; the 56th opposite the anus. The scales between have very obsolete spots of lighter, scarcely discernible. The sides are yellow, with black spots corresponding to the dorsal lines; indeed, there may be indistinctly discerned two or three lateral series of alternating blotches.

In larger specimens from the West, this tendency in the spots to aggregation is still more distinct. The back is crossed by these dotted lines of the number and relation indicated, at intervals of four or five scales; the spots on the intervening space being obsolete. These lines bifurcate at about the 9th outer row, the branches connecting with those contiguous, so as to form hexagons; and these extending towards the abdomen again, decussate on about the third outer row, thus enclosing two series of square, dark spots on each side. These lateral markings are, however, not very discernible, owing to the confusion produced by the greater number of yellow spots. On the edge of the abdomen are dark blotches, one opposite each dorsal dark space, the centres of the scutellæ being likewise blotched, but so as rather to alternate with those just mentioned.

The specimen represented on Plate VII was caught the 16th of May, between Cache creek and Red river. Total length $33\frac{1}{4}$ inches; tail $3\frac{3}{4}$ inches; abdominal scutellæ 224; subcaudals 49.

8. Ophibolus gentilis, B. & G.

Zoology, Pl. VIII.

Spec. char.—Muzzle more convex and acute than in *O. doliatus*. Body red, encircled by about 25 pairs of broad black rings enclosing a white ring: the white mottled with black on the sides. Black rings broader than in *O. doliatus*. Upper part of head entirely black.

Syn.—*Ophibolus gentilis*, B. & G. Cat. N. Amer. Rept. I, 1853, 90.

Desc.—Ground-color deep-red, encircled by 25 pairs of black rings, the 21st opposite the anus, each pair enclosing a third ring of white, the latter becoming yellowish by immersion in alcohol. The black rings are conspicuously broader above, the three crossing eight scales on the vertebral row anteriorly, and towards the anus about five. Anteriorly the intervals between successive pairs consist of about five scales, posteriorly only of two or three, thus diminishing considerably. The black rings contract as they descend, those of each pair receding slightly from each other, so as to cause the yellow portion to expand about one scale. The black rings are continuous on the abdomen, those of contiguous pairs (not of the same pair) sometimes with their intervening spaces black. The scales in the white rings are always more or less mottled with black, especially along the sides of the body, this mottling being very rarely observable on the red portion. The anterior black ring of the first pair is extended so as to cover the whole head above, except the very tip; the white ring behind it involves the extreme tip of the occipitals.

A large specimen is much duskier in its colors. The black rings extend on the back so that the contiguous rings of adjacent pairs run into each other. There are 28 pairs of rings, the 25th opposite the anus.

Dorsal row of scales, 21; abdominal scutellæ, 198; subcaudal ones, 45. Total length, 20 inches; length of tail, 2¾ inches. Plate VIII represents the largest of two specimens, caught June 14, near Sweetwater creek.

VII. MASTICOPHIS, B. & G.

The prominent feature of this genus consists in a very slender and elongated tail, forming one-third or one-fourth of the length. It bears a close relationship to the black snakes (*Bascanion*), from which it differs chiefly in the structure of the plates on the upper jaw. The scales are smooth and disposed in fifteen or seventeen dorsal rows. The preanal scutella is divided. The vertical plate is long and narrow. There are two anteorbitals and two postorbitals, these resting against the fifth labial.

9. MASTICOPHIS FLAVIGULARIS, B. & G.

SPEC. CHAR.—Light dull yellow, tinged with brown above. Beneath, two longitudinal series of blotches distinct anteriorly. In alcohol, and especially when the epidermis is removed, the whole animal appears of a soiled white.

SYN.—*Psammophis flavigularis*, Hallow. Proc. Acad. Nat. Sc. Philad. VI, 1852, 178.

Masticophis flavigularis, B. & G. Cat. N. Amer. Rept. I, 1853, 99.

DESC.—Size very large. Vertical plate broad before, tapering to the middle, where it is about half as wide as anteriorly, thence it runs nearly parallel. Vertical rather shorter than occipitals. Greatest breadth across superciliaries less than half the length of the portion covered by plates. Occipitals moderate. Centre of eye considerably anterior to the centre of commissure, over the junction of the 4th and 5th labials. Labials 8 above, increasing in size to the 5th, which is elongated vertically, the 7th elongate and largest. The 5th forms part of the inferior and posterior wall of the orbit, as in all the species of the genus, resting above against the lower postorbital, with which the 6th labial is not in contact. Dorsal scales broader than in *Bascanion constrictor*, their sides perfectly straight, slightly truncate, with the corners rounded. Exterior row largest, rest gradually diminishing. Scales on the tail widely truncate.

The general color, both above and below, may be described as a dull straw-yellow, tinged with light olivaceous brown above. This latter tint exists in the form of a shading on the centres and towards the tips of the scales, leaving the bases yellowish. The proportion of brown increases towards the back, and in older specimens sometimes suppresses the yellow. In all instances a darker shade is seen towards the tip of each scale. The skin between the scales is yellowish. The scutellæ anteriorly exhibit each two rather large brownish blotches, one on each side of the median line, constituting two rows on the abdomen, which fade out posteriorly. Sometimes the series are not discernible, the blotches spreading so as to constitute a dark shade to the margins and exterior edges of the scutellæ. The posterior portions of the plates under, and on the sides of the head, are similarly blotched; the same tendency being observable on the posterior edges of the plates on the top of the head, by the deeper shade of the olivaceous brown there prevalent. Anteorbitals yellow. One specimen was procured $57\frac{1}{2}$ inches long. Abdominal scutellæ $191+2$. The tip of the tail is missing.

In smaller specimens the blotching beneath is rather more decided. In addition to the colors described, the back is crossed by indistinct bars of darker, eight or nine scales wide and half a scale long. This color is also seen on the skin between the scales under the dark bars, where the bases of the scales themselves are darker instead of light. There is a tendency towards stripes on the side: first one of light brown, on the outer edge of the abdomen; then an interrupted yellow one at the junction of the abdominal scutellæ and outer scales; then brown again through the centres of the rows. This, however, is not very conspicuous. Sometimes the dark shades on the sides are tinged with reddish. The obsolete transverse bars are seen at intervals of one or two scales.

This species may prove to be the *Coluber testaceus* of Say. A specimen from Fort Webster, or the Copper Mines, collected by the U. S. and Mexican Boundary Commission, shows the stripes on the sides much more distinctly, running through all the dorsal rows anteriorly, and crossed by the indistinct bars already referred to. The contrast between the dark chestnut-brown spots on each side, and its deeper centre, with the clear yellow of the edges, is very distinct. Beneath yellow, with the blotches reduced to mere dull spots.

VIII. LEPTOPHIS, Bell.

The body is elongated and very slender, the tail forming more than the third of the entire length. The scales are disposed in 17 dorsal rows and keeled, except the two outer rows, which are smooth. The nostril is situated in the middle of a single plate. The eyes are large. The preanal scutella is bifid or divided. The color uniformly green.

10. LEPTOPHIS MAJALIS, B. & G.
Zoology, Pl. IX.

SPEC. CHAR.—Reddish green above, yellowish white beneath. Body proportionally stouter and tail shorter than in *L. æstivus*. Snout and whole head, including vertical, longer than in latter species. Dorsal scales in 17 rows.

SYN.—*Leptophis majalis*, B. & G. Cat. N. Amer. Rept. I, 1853, 107.

DESC.—Head more pointed, broader on the temporal region, and more tapering on the snout than in *L. æstivus*. Vertical plate subhexagonal, broader, and postfrontals proportionally larger in comparison with the prefrontals, than in *L. æstivus*. Occipitals maintaining more their width posteriorly, obtuse-angled behind. Nasal more elongated; loral smaller, and longer than high. Two large temporal shields and a few small ones behind. Scales strongly carinated, except the outer row, which is perfectly smooth, and the second row, which is but slightly carinated. The scales of both of these rows are broader than the rest.

Total length, $28\frac{1}{2}$ inches; length of tail, $9\frac{3}{4}$ inches. Abdominal scutellæ, 163+1; subcaudal, 111. The specimen figured on Plate IX was caught on the 13th of July at the head of Cache creek, near old Witchita village.

LIZARDS.

Six species of saurians, or lizards, belonging to six different genera, were collected during the exploration of Red river. One of them has proved to be new to science; two were recently described for the first time, whilst the three others have been long known to herpetologists.

I. PHRYNOSOMA, Wiegm.

This genus, including the so-called horned toads or horned frogs, more properly horned lizards, is recognizable by a depressed, broad, and subelliptical body, covered above with irregular scales, the majority very small, others quite large, pyramidal, raised above the surface of the skin, and scattered all over the back, sides, and tail. The head is subtriangular and provided with powerful spines or horns, giving to it rather a formidable appearance, although all the species of this genus are perfectly inoffensive. There are external auditive apertures as in most of the lizard tribe.

In a monograph of the genus appended to Stansbury's Exploration of the Valley of the Great Salt Lake of Utah, six species are described as indigenous to North America; another has since been added to the list. The single species collected is the most abundant of the genus.

1. PHRYNOSOMA CORNUTUM, Gray.

SPEC. CHAR.—Nostrils situated within the internal margin of the superciliary ridge; occipital and temporal spines longer and more acute than in *Ph. orbiculare;* a double row of pyramidal scales on the sides of the abdomen; scales on the inferior surface of the head small and slightly keeled, of a general uniformity, except one row on each side, somewhat larger, pyramidal, acute, slightly raised, and directed outwards and backwards; a series of very large inframaxillary plates, sharp on their outer edge, the posterior one of which is transformed into a spine. The plates lining the margin of the jaws are not prominent. The scales of the belly are proportionally small, subquadrangular, keeled, and posteriorly very acute; femoral pores undeveloped, or rudimentary in the female.

Syn.—*Phrynosoma cornutum*, Gray, Syn. Rept. in Griff. Anim. Kingd. IX, 1831, 45. Holbr. N. Amer. Herp. II, 1842, 87. Pl. xi.—Girard in Stansbury's Expl. Val. Great Salt Lake, 1852, 360. Pl. viii, fig. 1—6.

Agama cornuta, Harl. Med. and Phys. Res. 1835, 141. Plate, figs. 1 and 2.

Phrynosoma Harlani, Wiegm. Herp. Mex. 1834, 54.—Dum. and Bibr. Erp. gen. IV, 1837, 314.

Obs.—The color of this species has been well described by Dr. Holbrook. We may add that the ground-color above in some individuals is of a variable shade of ferruginous red—a tint sometimes seen on the inferior surface of the body. The belly is either unicolor, or else spotted as in *P. orbiculare*. Numerous specimens of this species were collected during the exploration of Red river; some on the prairie between Camps 2 and 3; others between Camps 6 and 7, and at Camp 7 also; others still on the south fork of Red river, and several other localities.

II. CROTAPHYTUS, Holbr.

Noticed for the first time during Major Long's expedition. The type of this genus was briefly described by Say in the second volume of Long's Narrative, and there called *Agama collaris*, in allusion to the very striking feature of bearing a double black sub-crescentic band on the sides of the neck. The genus *Crotaphytus* was first established by Dr. Holbrook, and is characterized by the presence of small, polygonal plates on the whole surface of the head. The odd occipital plate itself is inconspicuous; the auditive apertures are very broadly open. Teeth are found on the jaws and palate. There is a fold of the skin under the throat; the head is large and sub-triangular; the body covered with minute scales; and the tail very long and tapering. Femoral pores present.

This genus now includes four North American species; three we have lately described under the names of *C. Wislizenii*, from New Mexico, *C. Gambelii*, from California, and *C. dorsalis*, from the desert of Colorado; the fourth is the following:

2. Crotaphytus collaris, Holbr.

Spec. char.—Tail conical, very long and tapering; head large, sub-triangular, rounded at the snout; two subcrescentic black bars, margined with white on each side of the neck; the largest extends from the origin of the fore-legs to near the dorsal line; the second of these black bars is smaller, and situated between the latter and the head.

Syn.—*Crotaphytus collaris*, Holbr. N. Amer. Herp. II, 1842, 79. Pl. x.

Agama collaris, Say, in Long's Exp. Rocky Mts. II, 1823, 252.—Harl. Med. and Phys. Res. 1835, 142.

Obs.—The specimens on hand exhibit several varieties of coloration worthy of being noticed: thus two specimens from Gypsum Bluffs, on Red river—a rocky locality—present a green ground-color above, with large blue patches and bright yellow spots; underneath light-green, almost uniform, except under the head, which is deeper and provided with blue, irregularly elongated spots; another specimen from the same locality has brown as the predominating tint. Light-brown stripes are seen on the legs and tail; similar spots on the body and head; four rows of red spots on the back; belly light-brown; light reddish-brown under the tail and feet.

Specimens from the head of the south fork of Red river have either a bluish-gray back, with white spots, a bluish-white belly, and the inferior surface of fore-legs reddish, or else the back is yellow and green.

The above memoranda, on the coloration, were taken on the spot by Capt. Marcy. The general distribution of color appears to indicate sexual differences: thus all the specimens before us in which the spots have a tendency to arrange themselves in transverse bands, or even where transverse narrow bands take the place of the spots, have proved to be females. The ground-color, however, varies in both sexes.

III. HOLBROOKIA, Girard.

The genus *Holbrookia* bears a striking resemblance to the one just described; it has the same general form, the same sub-triangular head, covered with small polygonal plates, a fold under the throat, small scales on the back, and femoral pores. The tail is perhaps smaller in proportion to the size of the body. The absence of an external auditive aperture will, however, at once characterize it generically from all its allies. The absence of teeth on the palatine bones is another organic character by which the genus *Holbrookia* can be distinguished from *Crotaphytus*. From *Homalosaurus* it differs only by the absence of an external auditive aperture.

The species upon which the genus was originally based is the one collected by the expedition.

Three other species were found in Texas, and described by us under the names of *H. affinis, propinqua*, and *texana*, (see Proceedings of the Academy of Natural Sciences of Philadelphia, August, 1852).

3. HOLBROOKIA MACULATA, Girard.

SPEC. CHAR.—Above light-brown, with two dorsal series of irregularly crescent-shaped black spots convex posteriorly, and provided with an olivaceous margin; flanks with small crowded yellowish or reddish spots; two, occasionally three, deep-bluish black spots on the sides of the abdomen; beneath unicolor, either of a soiled white or yellow tint; sometimes irregular bluish vittæ under the head.

SYN.—*Holbrookia maculata*, Girard, Proc. Amer. Assoc. Adv. of Sc. IV, (1850), 1851, 201; and in Stansbury's Expl. Valley of Great Salt Lake, 1852, 342. Plate vi, fig. 1—3.

OBS.—A full description, as well as a figure of this species, may be found in the Report of Captain Howard Stansbury on the Valley of the Great Salt Lake of Utah; rendering it unnecessary to reproduce either here.

Numerous specimens were collected on the Canadian river and surrounding localities.

IV. SCELOPORUS, Wiegm.

The genus has the general appearance of *Holbrookia*, but is provided with large auditive apertures, large imbricated and carinated scales on the back in most instances; and smooth scales on the belly. The subguttural fold of the former, however, is not to be seen here and on the surface of the head; the plates, though small, are larger, especially the occipital. There are no teeth on the palate.

Most of the species of this genus are Mexican: one is common in the United States, and known as the brown or fence lizard. A second species was discovered by Captain Stansbury in the valley of the Great Salt Lake. Another species inhabits the western States, and a fourth is peculiar to Oregon.

4. Sceloporus consobrinus, B. & G.

ZOOLOGY, Pl. X, Figs. 5-12.

SPEC. CHAR.—Ground color above brownish, with a series of small black spots, eight or ten in number, on each side of the dorsal line. A yellowish stripe outside of the spots, and a black band beneath the stripe. A greenish area between the black band above and the elongated blue patch on the sides of the abdomen. Beneath, greenish blue.

DESC.—This species bears a close relationship to *S. graciosus*, from the valley of the Great Salt Lake of Utah, in the description of which the remarkably large size of the dorsal scales was mentioned as constituting one of its most distinguishing features, when compared to *S. scalaris*. In the present species the dorsal scales are proportionally still larger than in *S. graciosus*. Its body and head are also more slender and narrower. The tail is more tapering and elongated, and constitutes almost three-fifths of the total length. In coloration the differences between the two species are very striking.

The head is subelliptical, depressed, declive towards the snout, which is rather pointed. The superciliary region is but slightly raised above the plane of the vertex. The rostral plate is subtriangular, very low, and elongated transversely. The nostrils are almost circular, situated in the middle of a small plate, separated from the rostral by two small intervening ones. There are ten or twelve internasal and very small plates, and nine somewhat larger frontals, the middle one the largest. There are two verticals (or frontals), the anterior one the largest. The occipital is large and pentagonal, surrounded by four or six smaller plates, two anterior of medium size, contiguous to the postvertical, two lateral, larger and triangular, exteriorly to which two smaller ones may be observed. There are three or four subhexagonal, transversely elongated, plates on the superciliary region, surrounded internally by one row and externally by two rows of minute plates. The superciliary edge is formed by five sharp and imbricated thin plates; it is continued in the shape of a ridge to the nostril by means of two sharply keeled plates. The suborbitals are two in number, the posterior one much the longest. There is a small loral.

The plates which line the jaws are subquadrangular, very narrow and elongated, four above and five below, the latter considerably larger. Above the series of plates of the upper jaw, and between the suborbitals, two series of small and irregular plates may be observed. Four or five inframaxillary plates constitute a series on each side of the lower surface of the head, joined anteriorly by the subpentagonal symphysal plate. Between the inframaxillary series and the series lining the lower jaw exists a series of four or five elongated and small plates.

The auditive aperture, which is proportionally large, is oval, and almost vertical in its longest diameter. At its anterior margin may be seen two or three scales, larger and more pointed than those on the temporal region. Behind the auditive aperture, and situated obliquely on the neck, is a slight fold of the skin.

The neck is somewhat contracted, the body slender and depressed, with the back, however, slightly arched, and the belly flat. The tail, as already mentioned, is quite long and slender, depressed at its base, and hence conical towards the tip.

The fore-legs, when stretched backwards along the sides, extend nearly to the groin, while the hind-legs, when brought forwards, reach almost to the ear.

The scales are imbricated and keeled on the back and sides. Their general shape is that of a lozenge, terminated posteriorly by an acute spine. There are ten longitudinal rows along the back, with five on each side, which are somewhat oblique, and smaller. Underneath, the scales are smooth, posteriorly tricuspid on the belly, whilst under the head and throat they have but two posterior spines. The scales below as well as above the fore-legs are keeled. Those on the upper part of the hind-legs are also keeled, whilst on the thighs they are smooth. The fingers and toes are surrounded with carinated scales to their very tips.

The femoral pores, thirteen or fourteen in number on each side, are conspicuous and situated in the middle of one single small plate.

The black spots in the series along the back are comparatively small, and separated from each other by a space greater than their diameters. The yellow stripe extends from the origin of the neck to beyond the anus, the black from above the shoulder to the groin. The blue patch is elongated and narrow, terminated posteriorly by a black stripe which runs for a little distance along the thigh. There is an elongated black spot on the shoulder.

One specimen was collected on the 6th of June.

Plate X, fig. 5, represents the species in profile and of the natural size.

Fig. 6. The side of the head enlarged, to show more distinctly the structure of its plates.
Fig. 7. Head from above, enlarged in the same proportion as fig. 6.
Fig. 8. Head from below, enlarged.
Fig. 9. The right arm and fingers, seen from below.
Fig. 10. The right leg and toes, seen from below.
Fig. 11. Dorsal scales, enlarged four times.
Fig. 12. Scales from the belly, slightly enlarged.

V. CNEMIDOPHORUS, Wagl.

This genus is characterized by a bifid tongue; a double transverse fold of skin under the throat; teeth on the palate; maxillary teeth compressed, the posterior one tricuspid; femoral pores; broad plates under the thighs; fingers not carinated underneath; and a subcylindrical, very long and tapering tail. The body above is covered with minutely crowded scales; whilst on the belly there are eight longitudinal rows of subquadrangular, transversely elongated plates, or scutellæ. On the tail the scales are quite large and very conspicuous, strongly carinated and constituting circular rows or whorls.

The explorations of the last few years in Texas and New Mexico have brought to light several other species of the genus *Cnemidophorus*, all provided with eight longitudinal rows of abdominal scutellæ. These are *C. gracilis*, from the desert of the Colorado; *C. perplexus*, from the upper valley of the Rio Grande; *C. gularis*, *C. Grahamii*, and *C. marmoratus*, from different localities in Texas.

The discovery of *C. gularis* in Arkansas is an interesting fact in regard to its geographical distribution.

5. CNEMIDOPHORUS GULARIS, B. & G.
ZOOLOGY, Pl. X, fig. 1–4.

SPEC. CHAR.—Ground color brownish, with six longitudinal stripes, green or yellow; beneath yellowish white, unicolor. Scales on the subguttural fold quite large and conspicuous in proportion to those in other species.

Syn.—*Cnemidophorus gularis*, B. & G. Proc. Acad. Nat. Sci. Philad., vi, 1852, 128.

Desc.—This species is very closely allied to *C. sexlineatus*, having, like the latter, six longitudinal stripes, three on each side of the body, running from head to some distance along the tail. It has, also, the same general form; but on a close comparison it will soon be observed that the body is proportionally shorter, the limbs more developed, whilst the scales on the back appear to be actually larger. The head is proportionally smaller and narrower. But the most striking organic character consists in the presence of somewhat large and conspicuous scales on the margin of the subguttural fold of the skin. The following indications of color are derived from the notes of the Expedition: The upper surface of the head is reddish brown; three longitudinal yellow or greenish stripes extending from the head to the origin of the tail; the middle stripe on each side may be followed on the tail to a considerable distance. The dorsal space between the two uppermost stripes on each side is brown, or reddish brown, like the head above. The space between the uppermost and middle stripes is of a deep black, and extends from the upper angle of the orbit down to a certain distance along the tail. The space between the middle and lower stripes, and between the latter and the abdominal scutellæ, is green, or greenish brown. The legs are brownish red, and the belly white or bluish white. The tail underneath is yellowish red; above, brownish, or reddish brown.

Specimens were collected on the 5th and 6th of June.

Plate X, fig. 1, represents *Cnemidophorus gularis* of natural size.

Fig. 2. Head seen from above, to exhibit the plates.

Fig. 3 shows the scales on the subguttural fold and the hand from beneath, as well as the submaxillary plates.

Fig. 4 represents the femoral pores, the preanal plates, and also the plates at the inferior surface of the hind-legs, and the lower surface of the feet.

VI. LYGOSOMA, Gray.

This genus includes small scincoid lizards, the nostrils of which open in one single plate, the nasal. The supranasals are wanting. The palate is without teeth, and provided with a triangular notch situated far back. The scales, broader than long, are all smooth.

All the species of *Lygosoma* belong to the Old World, except the one here mentioned.

6. LYGOSOMA LATERALIS, Dum. and B.

SPEC. CHAR.—Upper part of head and body chestnut-brown; a black lateral band extending from the snout across the eye to a considerable distance along the tail. Flanks grayish-brown, with longitudinal indistinct, darker, interrupted vittæ. Abdomen yellowish, and tail beneath bluish; circumference of scales mottled with gray. Tail longer than the body. Limbs very small.

SYN.—*Scincus lateralis*, Say, in Long's Exp. Rock. Mts. II, 1823, 324.—Harl. Jour. Acad. Nat. Sc., V, 1827, 221, and VI, 1829, 12.—Holbr. N. Amer. Herp., first ed., I, 1836, 71. Pl. viii.

Scincus unicolor, Harl. Journ. Acad. Nat. Sc. Philad., V. i, 1825, 156.

Tiliqua lateralis, Gray, Syn. Rept., in Griff. Anim. Kingd., Cuv. IX, 1831, 70.

Lygosoma lateralis, Dum. and B., Erp. gen. V, 1839, 719. Holb. N. Amer. Herp, second ed., II, 1842, 133. Pl. xix.

This small and graceful species appears to be spread over a large portion of the United States. It is always met with running on the surface of the ground in forests, among dead leaves, never ascending either trees or shrubs like many other lizards.

The body is sub-quadrangular, the head continuous with it, and, like it, flattened above. The tail is sub-circular, tapering into a point. The plates of the head correspond with the descriptions which we have before us, except that the frontonasals are not contiguous, but separated by a small odd plate directly in advance of the vertical (sometimes called frontal). But this peculiarity of structure is not indicative of any specific difference.

The auditive apertures are large, circular, and their margin simple. The fore-legs, when extended forward, reach the eye. The hind-legs are a little longer and stouter than the fore-legs. The scales are perfectly smooth, uniform above and below, and disposed in thirty longitudinal rows around the body. The two middle preanal scutellæ considerably larger.

One specimen was procured near the mouth of Cache creek, on the 16th of May.

BATRACHIANS.

Of this order of reptiles only two species were procured—a toad and a frog.

1. BUFO COGNATUS, Say.

ZOOLOGY, Pl. XI.

SPEC. CHAR.—Greenish brown above, with a lighter yellowish dorsal line. Patches of blackish-brown scattered over the sides and legs. Beneath unicolor of a dingy yellow. Head short, groove on its upper surface, not extending to the anterior rim of the eye.

SYN.—*Bufo cognatus*, Say, in Long's Exp. to Rock. Mts. II, 1823, 190.

OBS.—It is not without hesitation that we have referred the present species to *Bufo cognatus;* the description of Say as cited is exceedingly brief, applying almost equally well to several allied species. The colors of our specimen vary considerably from the *B. cognatus* as described by Say; but the characters of the groove of the crown agree better. The mark of "head with a groove which hardly extends anteriorly to the line of the anterior canthus of the eye," although not strictly in accordance with our species, may, with some allowance, be made to answer to it. It is much to be regretted that the original specimen of Say was destroyed in the conflagration of the Philadelphia Museum, and thus all hopes of identification are lost. If, however, further explorations in Arkansas should yield many additional specimens, all differing as much as the present from Say's description, it will become necessary to assign a new name to it, especially if the true *B. cognatus* be at the same time detected.

DESCRIPTION.—The head is very short, the snout obtuse and truncate, with the nostrils subterminal. Upper surface of head grooved; groove subelliptical and short, not extending anteriorly to the anterior rim of the eye (fig. 2). The superciliary ridges thicken from before backwards, extending to the tympanum in passing obliquely behind the eyes, and in contact also with the parotid glands, which are subovoidal and of medium size. Tympanum rather small, subelliptical; its longest diameter almost vertical. The fore and hind legs are well proportioned to the size of the body. The under surface of the hand is provided with small crowded tubercles, a more conspicuous and a larger one at the articulations of the fingers; the fingers themselves are depressed or flattened. A larger disc-like knob is observed on the middle and at the base of the hand (fig. 3).

The toes (fig. 4) are but slightly webbed, and, like the fingers, depressed. The fourth is conspicuously the longest, and the third a little longer than the fifth. The under surface of the feet (fig. 5) is covered with smaller tubercles than those of the hands. A large spade-like process exists at the base of the first or inner toe, exteriorly to which, and at the base still of the metatarsus, is a small knob-like tubercle. The body is thickly covered with papillæ, with some large ones more conspicuous along the sides of the back; on the flanks they are smaller, similar to those of the intervening spaces on the back; on the abdomen the papillae are smaller still; upper part of hands and feet minutely granulated. The snout alone is smooth.

The dark patches scattered over the upper part of the animal are margined with a light yellowish line. Sinuating yellowish lines may be observed on the sides of the belly, or flanks and legs. A rather large spot is seen beneath the eye, and another in advance and beneath the tympanum near the angle of the mouth.

One specimen procured near the Water-hole between Camps 6 and 7.

Plate XI, fig. 1, represents *Bufo cognatus* of natural size.

Fig. 2. The head from above.

Fig. 3. Left hand seen from below.

Fig. 4. Right foot from above.

Fig. 5. Right foot from below.

2. RANA PIPIENS, Latr.—Bullfrog.

SPEC. CHAR.—Toes webbed to their extremity, fourth toe one fourth longer than the third and fifth. An elongated tubercle at the base of the first toe; sub-articular tubercles of fingers and toes but slightly developed. Vomerine teeth on two rounded and separated elevations situated between the internal nostrils. Diameter of tympanum (in the specimen before us) greater than the diameter of the eye.

SYN.—*Rana pipiens*, Latr. Hist. Nat. Rept. II, 1802, 153. Harl. Amer. Jour. Sc. X, 62. Med. & Phys. Res., 1835, 101; and Jour. Acad. Nat. Sc. Philad. V, 1827, 335. Holbr. N. Amer. Herp. IV, 1842, 77. Pl. xviii.

Rana mugiens, Merr. Tent. Syst. Amph. 1820, 175. Dum. & B. Erp. gen. VIII, 1841, 370.

The bullfrog is quite a common animal in the United States, though its northern, western, and southern limits are not yet accurately known.

A large specimen was found in a cold spring near the head of the south fork of Cache creek, in the Witchita mountains. The upper parts of body and limbs are covered with warty eminences, more crowded on the body. These warts are perfectly smooth, like the skin itself. The ground-color is greenish brown above, with crowded deep brown or blackish spots. Beneath, dull yellow, with clouded bluish patches. The lower surface of the feet has the same marmorated appearance as the back. The jaws and snout are greenish brown, and perfectly smooth.

The specimen before us is remarkable for the size of its tympanum, which is much larger than the eye.

FISHES.

BY S. F. BAIRD AND C. GIRARD.

1. POMOTIS LONGULUS, B. & G.
ZOOLOGY, Pl. XII.

SPEC. CHAR.—General form elongated. Opercular flap rather small and entirely black. Twenty-seven to twenty-nine rows of scales across the line of greatest depth of body, and about thirteen rows on the tail. Fifty-two scales in the lateral line.

SYN.—*Pomotis longulus*, B. & G. Proc. Acad. Nat. Sc. Philad. VI, 1853, 391.

DESCRIPTION.—The body is very much compressed, and more elongated than usual in the genus *Pomotis*—so much so, indeed, as to resemble *Grystes* even more than *Centrarchus*. The head constitutes a little less than the third of the total length, including the caudal fin; it is subconical, with a little depression upon the middle of the skull. The eyes are large and circular, and their diameter is contained five times in the length of the head measured from the tip of the snout to the extremity of the opercular flap. The posterior extremity of the maxillary reaches a point opposite the middle of the pupil. The cheeks are densely covered with small and imbricated scales. The largest scales are on the opercular apparatus (the preopercular excepted), where they are also imbricated. The opercular is subtriangular; its upper angles rounded, and the posterior one terminated by a membranous and rather small flap, entirely black. The subopercular extends along the inferior edge of the opercular, tapering slightly upwards. The interopercular forms a regular curve immediately beneath the preopercular, and is covered with one row of scales, there being a double row of these upon the subopercular.

The dorsal fin is rather low, especially its spiny portion. Its anterior margin is exactly opposite to the opercular flap. There are ten spiny rays and nine soft ones, the last being double and the shortest. The first, second, third, fourth, fifth, and sixth rays increase gradually in length in the order enumerated;.the eighth is equal to the sixth; the ninth is the longest. They all (the soft rays) bifurcate from their middle, and then again subdivide from four-fifths of their length to the tip. The caudal fin is subcrescentic posteriorly; its angles are rounded; its length contained five times and a half in that of the body and head together. The central rays bifurcate three times upon their length. There are seventeen rays in all, with a few rudimentary ones. The anal is well developed; its three anterior spiny rays are the shortest, and not very conspicuous. The eight remaining ones are soft and articulated; similar in structure to those of the dorsal fin. The ventrals are inserted behind the base of the pectorals; their tip, when bent backwards, reaching the anus, which is situated a quarter of an inch in advance of the anterior margin of the anal fin. This is subtriangular, posteriorly subtruncated, composed of an anterior spiny ray and five soft and articulated ones, which bifurcate twice. The pectoral extends backwards as far as the ventrals. Its rays, fifteen in number, are all soft and very slender, bifurcating twice. Only thirteen of these rays are well developed. The formula of the fins is as follows:

D X. 9 + 1; A III. 8; C 2. I. 8. 7. I. 1; V I. 5; P 15.

The scales are of medium size, longer than high, truncated anteriorly, rounded posteriorly, and finely denticulated, as seen in fig. 4. The lateral line does not extend beyond the insertion of the rays of the caudal, the base of which is covered with scales irregularly disposed. The smallest scales are observed under the head, upon the throat; the largest on the peduncle of the tail.

The color is not sufficiently preserved in the single specimen collected to admit of description. Traces of irregular lines are, however, visible upon the cheeks and opercular apparatus.

Found in Otter creek, Arkansas.

Plate XII, fig. 1. *Pomotis longulus*, in profile and of the size of life.

Fig. 2. A dorsal scale taken on the middle of the back, above the lateral line.

Fig. 3. A scale from the lateral line, exhibiting the mucous tube.

Fig. 4. A scale from the sides of the abdomen, below the lateral line.

Figs. 2–4 are magnified twelve times.

2. POMOTIS BREVICEPS, B. & G.

ZOOLOGY, Pl. XIII.

SPEC. CHAR.—General form short and stout, subelliptical; opercular flap very much developed and directed upwards, black with a lighter margin. Twenty-four rows of scales across the line of greatest depth, and twelve rows on the peduncle of the tail. Thirty-seven scales on the lateral line.

Syn.—*Pomotis breviceps*, B. & G. Proc. Acad. Nat. Sc. Philad. VI, 1853, 309.

Description.—The abbreviated head constitutes two-sevenths of the entire length, the caudal fin included. Middle of the cranium slightly depressed. Nostrils nearer to the eye than to the tip of the snout. Eyes of medium size and circular; their diameter is contained four times only in the length of the head from the snout to the base of insertion of the opercular flap, which is longer than a diameter of the eye. The mouth is proportionally small; the posterior extremity of the maxillary not extending as far back as the middle of the pupil. The teeth are slenderer and more conspicuous than in *Pomotis longulus*. Cheeks covered with scales, but slightly smaller than those on the opercular apparatus. The opercular bones have the same general shape as in *Pomotis longulus;* but the opercular flap is very much developed, longer than broad, and rounded posteriorly. The structure of this flap is somewhat similar to a fin; slender and simple rays being distinctly visible through the membrane.

The dorsal fin commences in a line above the base of the opercular flap. Its spinous portion is almost as elevated as the soft one. There are eleven spinous rays, the first and second smallest; and ten articulated or soft ones, occupying a little more than half the space as the spines. The structure of the soft rays and their relative length are much as in *Pomotis longulus;* they bifurcate from the middle of their length, and subdivide again upon their extremity. The caudal is subcrescentic posteriorly, and its angles rounded. It is composed of seventeen well developed rays, and a few rudimentary ones. The central rays bifurcate three times. The anal fin is composed of nine soft rays, one more than in *P. longulus*, and three spinous ones; the second and third almost equal in length. They bifurcate and divide in the same manner as the dorsal. The insertion of the ventrals is immediately behind the base of the pectorals. Their shape is triangular, and when bent backwards their tip extends to the anterior margin of the anal fin, thus overlapping the anus, which is situated as in *P. longulus*, about a quarter of an inch in advance of the anterior margin of the anal fin. The ventrals are composed of one spinous and five articulated rays, which bifurcate three times. The insertion of the pectorals is subcrescentic; the tip of these fins reaches about as far back as do the ventrals. The rays, fourteen in number (thirteen of them well developed), are slender and show traces of a bifurcation of the third degree upon their extremity.

D XI. 10; A III. 9; C 2. I. 8. 7. I. 2; V I. 5; P 14.

The scales are proportionally large, higher than long, subtruncated anteriorly, and rounded posteriorly with minute denticulations. Scales irregularly disposed, exist on the base of the caudal. The largest scales are seen on the middle of the flanks, and the smallest upon the subthoracic region. The ground-color appears to have been of a uniform reddish brown. The opercular flap is deep black, margined with a lighter line, the hue of which is not preserved. There are several sinuous irregular lines upon the cheeks and opercular apparatus.

This species has a general resemblance to *Pomotis nietidus*, Kirt., but may easily be distinguished by prominent characters.

One specimen was caught in Otter creek, Arkansas.

Plate XIII, *Pomotis breviceps* of natural size.

Fig. 2. A dorsal scale.

Fig. 3. A scale from the lateral line.

Fig. 4. A scale from the sides of the belly.

Figs. 2–4 are enlarged twelve times.

3. Leuciscus vigilax, B. & G.*

ZOOLOGY, Pl. XIV, figs. 1–4.

SPEC. CHAR.—Subfusiform. Dorsal fin longer than high. Sixteen rows of scales across the line of greatest depth, and eight on the peduncle of the tail. Thirty-eight to forty scales in the lateral line, which runs through the middle of the sides, slightly bent downwards on the abdomen.

SYN.—*Ceratichthys vigilax*, B. & G. Proc. Acad. Nat. Sc. Philad. VI, 1853, 391.

DESCRIPTION.—Body subfusiform, compressed. The head forms one-fifth of the entire length from the snout to the tip of the caudal fin; it is contained three times in the length of the body, the caudal fin being about one-fifth of the entire length. The head itself has the shape of a truncated cone. The eyes are subelliptical; their longitudi-

---

* NOTE.—Owing to the immature state of the specimens, we have preferred returning this species to the genus *Leuciscus*. Although having a strong resemblance to *Ceratichthys*, as also to *Pimephales*, in the bluntness of the snout, the inferior position of the mouth, and other characters, yet the specimen is almost too small to allow a final determination as to its generic character.

nal diameter being contained three times and a half in the length of the sides of the head. The mouth is rather small, its angle not extending to a point below the anterior rim of the eyes. The opercular bone is conspicuously large, and almost trapezoidal in shape. The subopercular and preopercular are comparatively small. The isthmus beneath is about three-tenths of an inch wide.

The dorsal fin is longer than high, and is composed of nine rays, bifurcated from about their middle; some of the median rays showing another subdivision upon their extremity. The caudal fin is forked; its angles are acute. It contains eighteen well developed rays, and several rudimentary ones above and below; the central ones bifurcate twice. The base of the caudal fin is considerably broader (higher) than the central portion of the peduncle of the tail. The anal fin is situated behind the dorsal, is higher than long, subtrapezoidal, and composed of eight bifurcated rays; the central ones subdivided towards their extremity. The ventrals are inserted very little behind the anterior margin of the dorsal; they are rather slender, posteriorly rounded, composed of eight bifurcated rays, the middle ones bifurcated towards their extremity; and when bent backwards the fin does not reach quite to the anus, which is situated immediately in advance of the anterior margin of the anal fin. The pectorals are slender; when bent backwards they do not reach the insertion of the ventrals. They are composed of fourteen bifurcated rays, the central ones subdividing at their last third. Formula:

$$D\ 9;\ A\ 8;\ C\ 3.\ I.\ 8.\ 8.\ I.\ 3;\ V\ 8;\ P\ 14.$$

The scales are proportionally large, a little higher than long, rounded at both extremities, more abruptly posteriorly. The lateral line runs along the middle of the side, slightly bent downwards on the abdomen.

The ground-color is yellowish brown; a blackish stripe composed of crowded dots follows the lateral line on the sides.

One specimen (immature) caught in Otter creek, Arkansas.

Plate XIV, fig. 1, represents *Leuciscus vigilax*, size of life.

Fig. 2. A dorsal scale.

Fig. 3. A scale from the lateral line.

Fig. 4. Abdominal scale.

Figs. 2–4 are enlarged twelve times.

4. LEUCISCUS BUBALINUS, B. & G.

ZOOLOGY, Pl. XIV, figs. 5–8.

SPEC. CHAR.—Compressed. Back arched. Tail slender. Dorsal fin higher than long. Ten rows of scales across the line of greatest depth, and five rows on the tail. The lateral line, which contains about thirty-six scales, runs below the middle of the flanks. Dorsal, caudal, anal, and ventral fins, well developed.

    Syn.—*Leuciscus bubalinus*, B. & G. Proc. Acad. Nat. Sc. Philad. VI, 1853, 391.

    Description.—The body much compressed, and rather short in appearance. Back considerably arched in advance of the dorsal, behind which the body tapers quite rapidly posteriorly, rendering the peduncle of the tail comparatively slender. The head is about one-fifth of the entire length. Eyes comparatively large and circular; their diameter contained three times and a half in the length of the head, one diameter intervening between the eye and the snout. The nostrils are nearer to the eyes than to the tip of the snout. The jaws are even, (the figure represents the lower one a little too short). The opercular apparatus is conspicuously developed, especially the opercular, which has the shape of an elongated quadrangle, slightly concave posteriorly, and slightly rounded inferiorly. The isthmus is quite small.

    The anterior margin of the dorsal fin corresponds to the middle of the distance between the snout and the base of the caudal fin. It is angular and higher than long, and composed of eight rays. The anal has the same length as the dorsal, but is not quite as high; it is composed of nine articulated rays and two minute spines at the anterior margin. The ventrals when bent backwards reach the anterior margin of the anal fin, consequently overlapping the anus situated close to the anal fin. They contain eight rays, all soft or articulated. The pectorals are comparatively small and slender, reaching the insertion of the ventrals when brought backwards. Their posterior margin is rounded; the rays eleven in number. In all the fins the rays are bifurcated, and the middle ones subdivided upon their length. Formula:

$$D\ 8;\ A\ II.\ 9;\ C\ 4.\ I.\ 9.\ 9.\ I.\ 3;\ V\ 8;\ P\ 11.$$

    The scales are large, higher than long, rounded anteriorly, subtruncated posteriorly. The lateral line forms a very open curve, convex downwards, and nearer to the insertion of the ventrals than to the base of dorsal.

    The ground-color is greyish; the hue is not preserved on the specimen.

    Caught, like the preceding, in Otter creek, Arkansas.

    Fig. 5 represents *Leuciscus bubalinus* the size of life, and apparently quite mature.

    Fig. 6. A dorsal scale.

    Fig. 7. Scale from the lateral line.

Fig. 8. Abdominal scale.

Figs. 6–8 are enlarged twelve times.

5. Leuciscus lutrensis, B. & G.

Zoology, Pl. XIV, figs. 9–12.

Spec. char.—Subfusiform, compressed. Insertion of ventrals in advance of dorsal. Twelve rows of scales across the line of greatest depth; six rows on the tail. About thirty-six in the lateral line, which is bent downwards on the abdomen and slightly broken in advance of the anal fin. Dorsal and anal fins well developed.

Syn.—*Leuciscus lutrensis*, B. & G. Proc. Acad. Nat. Sc. Philad. VI, 1853, 391.

Description.—The body is much compressed and subfusiform in general appearance, somewhat tapering from the posterior margin of the dorsal and anal fins to the caudal, the base of which is broader than the peduncle of the tail. The greatest depth is equal to the length of the sides of the head, which is contained three times and a half in the total length, the caudal fin included. The greatest thickness is nearly half of the depth. In general aspect it resembles *Leuciscus kentuckiensis* of Kirtland. The eyes are of medium size, subcircular; their diameter contained four times in the length of the sides of the head. The nostrils, situated towards the upper surface of the head, are nearer to the eyes than to the tip of the snout. The posterior extremity of the maxillary does not reach the vertical of the anterior rim of the orbit.

The upper and posterior margins of the opercular constitute a uniform curve, whilst the anterior and inferior margins are straight, forming a rather acute angle. The subopercular and interoperculars are comparatively small.

The anterior margin of the dorsal fin is situated on the middle of the distance between the snout and the base of the caudal; the fin itself is quadrangular, higher than long, and composed of eight rays, the last double, and the anterior rudimentary in close contact with the next. The anal is shaped somewhat like the dorsal; it has nine perfect rays, and an anterior rudimentary one. The caudal is deeply forked with acute angles, and shorter than the head. It is composed of nineteen well developed rays, and several rudimentary ones, above and below. The ventrals are posteriorly rounded, (a character not expressed in the figure,) composed of eight rays, and when bent backwards their tips reach the anus, which is situated immediately in advance of the anal fin. The pectorals are elongated, rather slender, rounded, and their tip not quite reaching the insertion of the ventrals. They are composed of eleven slender, bifurcated, but not subdivided, rays. The mèdian rays of the dorsal, caudal, anal, and ventrals, are subdivided for at least one-fourth of their length, the bifurcation beginning sometimes upon their middle. Formula:

$$D\ 8+1;\ A\ 1.\ 9;\ C\ 2.\ I.\ 9.\ 8.\ I.\ 1;\ V\ 8;\ P\ 11.$$

The scales are proportionally large, higher than long; anterior, superior, and inferior margins, uniformly rounded, posteriorly subtruncated. The lateral line is considerably bent down on the abdomen, and slightly broken in advance of the anal fin.

The ground-color, as preserved in alcohol, is dull bluish brown; the back is bluish; the dorsal fin yellowish brown; the caudal, pectorals, and ventrals, are reddish.

Several specimens were caught in Otter creek, Arkansas; the largest of which we have had figured.

Plate XIV, fig. 9, *Leuciscus lutrensis* size of life.

Fig. 10. A dorsal scale from the middle of the region between the dorsal fin and the lateral line.

Fig. 11. A scale of the lateral line taken beneath the dorsal fin.

Fig. 12. An abdominal scale taken beneath the lateral line, half way between the latter and the line of the belly.

SHELLS.

BY PROFESSOR C. B. ADAMS.

Amherst, Massachusetts,
December 1, 1852.

Dear Sir: I transmit herewith a list of the shells which were collected in Texas and upon Red river, by Captain Marcy;

And have the honor to remain, your obedient servant,

C. B. ADAMS.

President Hitchcock.

ACEPHALA.

1. Unio asperrimus, Lea.

The specimens have a great profusion of small tubercles on the umbones. One large specimen was taken May 22d, at the foot of the Witchita mountains; 3 mature and 5 young shells were taken in Otter creek, July 13th, near the same place; long. about 100° W.; lat. about 34° 35' N.

2. Unio, Sp. indet.

This may be a variety of the preceding; but with only one decayed specimen, we do not venture to describe it as a new species. It differs in having only a few large tubercles in two radiant series, of which one passes down the middle of the disc, and the other is on the posterior angle. A few small curved ridges proceed from this angle to the ligamentary margin. No label.

3. Unio tuberculatus, Barnes.

A single valve of a young specimen; no label.

4. UNIO ANODONTOIDES, Lea.

7 specimens were taken in Otter creek July 13th; one of them is 5 inches long, 2 to 3 inches high, and 1.6 inch wide.

5. UNIO PARVUS, Barnes.

2½ specimens were taken in Otter creek July 13th.

6. UNIO HYDIANUS, Lea.

2 specimens were taken near Fort Washita July 31st.

7. UNIO LÆVISSIMUS, Lea.

The specimen is for this species remarkably thick; the nacre is deeply colored with reddish-purple, and there are some fine radiating striæ behind the umbones. No label, but may have been taken in Otter creek, since it was in the same parcel with the next species.

8. UNIO GRACILIS, Barnes.

Several specimens were taken in Otter creek July 13th, and some July 15th, probably in a branch of Cache creek, a few miles west of Otter creek.

9. CYCLAS DISTORTA, Prime.

4½ specimens were taken in Otter creek July 13th.

GASTEROPODA.

10. BULIMUS LIQUABILIS, Reeve.

4 specimens (dead) were taken in Otter creek July 13th. This is the only terrestrial species in the collection.

11. PHYSA ANCILLARIA, Say.

3 specimens were taken in Otter creek July 13th. They are more shouldered than is usual, but not so much as the variety figured by Professor Haldeman, Monog. Physa, pl. 3, fig. 5.

12. PHYSA HETEROSTROPHA, Say.

12 specimens were taken in Otter creek July 13th.

13. LYMNÆA CAPERATA, Say.

2 specimens were taken May 16th, one day from Cache creek.

14. PLANORBIS LENTUS, Say.

Several specimens were taken with the preceding; also in Otter creek, July 13th.

Geographical Distribution.

Nos. 3, 11, 12, and 13 occur also through the western and eastern States. No. 8 has its northeastern limit in Lake Champlain. Although Ohio specimens of this species are easily distinguished from those of Lake Champlain, it is remarkable that these Texan shells cannot be distinguished from them. Nos. 5, 6, and 14 are southern species. No. 10 has hitherto been known only as a Texan shell. The remainder are western and southern species.

NOTICES OF ADDITIONAL SPECIES OF SHELLS: BY G. C. SHUMARD, M. D.

1. UNIO ANODONTOIDES, Lea.

Found in the Little Witchita, and in a small creek between Fort Washita and Fort Arbuckle; quite abundant, and the specimens very beautiful.

2. UNIO RUGOSUS, Barnes.

Occurs with the preceding species at all the localities above mentioned. The specimens are less ventricose than any we have seen from the Ohio basin. They approach more nearly to a variety brought by Prof. Litton from Red river of the north.

3. UNIO SILIQUOIDEUS, Barnes.

Found in a small creek between Fort Arbuckle and Fort Washita.

4. UNIO LÆVISSIMUS, Lea.

A few detached valves of this species were found on the banks of Otter creek.

5. ANODONTA IMBECILIS, Say.

Abundant, and very beautiful, in Beaver creek; more sparingly in a small creek between Fort Arbuckle and Fort Washita.

6. PLANORBIS TRIVOLVIS, Say.

Abundant in many of the streams from Fort Belknap to the sources of Red river.

7. PHYSA GYRINA, Say,

Beaver creek, Choctaw Nation.

8. SUCCINEA AVARA, Say.

Otter creek, Choctaw Nation.

9. CYCLAS PARTUMEIA, Say.

Otter and Beaver creeks.

10. BULIMUS DEALBATA, Say.

Texas.

ORTHOPTEROUS INSECTS.

BY CHARLES GIRARD.

I. DAIHINIA, Hald.

GEN. CHAR.—Body rather short, concave above, without any traces of wings; provided with short and robust limbs; second and third nts of tarsi, equal; antennæ long and filiform. A row of spines upon the under surface of the femora, more conspicuous in males than in females.

SYN.—*Daihinia*, Hald. Proc. Amer. Assoc. Adv. Sc. II, 1850, 346.

OBS.—The general aspect of this genus is that of *Phalangopsis*, from which it differs by having "shorter antennæ, shorter and more robust limbs." It approximates to *Stenopelmatus* by the structure of its tarsi, in which the second and third joints are equal.

Prof. Haldeman, who traced the distinction between *Daihinia* and *Phalangopsis*, proposed to consider the former as a mere sub-genus of the latter. But should the above character prove constant, they are sufficient to raise *Daihinia* to the rank of a genus; thus simplifying much the nomenclature.

Two species of this genus are known—the one herein described and figured, and *D. robusta*, Hald., an inhabitant of New Mexico.

1. DAIHINIA BREVIPES, Hald.
ZOOLOGY, Pl. XV, figs. 9–13.

SPEC. CHAR.—Dark brown, mottled with lighter shades; legs short and robust; tibiæ shorter than the femora, and strongly spinous; antennæ of medium development.

SYN.—*Phalangopsis (Daihinia) brevipes*, Hald. Proc. Amer. Assoc. Adv. Sc. II, 1850, 346.

DESCRIPTION.—The fact that in this species the tibiæ are shorter than the femora, contributes somewhat to impress upon it more strongly that character of the genus which consists in being provided with shorter limbs than in *Phalangopsis*. The surface of the body is generally smooth, but posteriorly, and particularly in the male, there are minute short spines, which give to that region a granulated appearance; these minute spines are especially crowded upon the margin of the segments or articulations. The femora are provided with spines above and below, stronger below, and more so in the male. The tibial spines are very much developed on the anterior and posterior tibiæ, much less on the medial ones; anteriorly they occupy the outer edge of the limbs, and answer fossorial purposes; posteriorly they constitute two rows, directed horizontally backwards, inclining a little downwards, the inner row being the strongest. The anterior and posterior tarsi are trimerous; the medial ones being tetramerous. The ovipositor is comparatively small; its length being less than the half of the length of the body, and provided beneath and towards the tip with from eight to ten small spines.

The ground-color is chestnut-brown, mottled above with lighter shades. The antennæ and spines are blackish.

Specimens were collected at the Camp No. 7, recorded as "yellowish-brown;" others on June 5th and 6th, said to be "yellowish-red."

Plate XV, fig. 9, represents the male *Daihinia brevipes* size of life.

Fig. 10 is the female, also the size of life.

Fig. 11, front view of the head of the female.

Fig. 12, a tarsus from above.

Fig. 13, a tarsus from below.

II. ANABRUS, Hald.

GEN. CHAR.—Body sub-cylindrical, thickest in the middle; without wings; antennæ almost as long as the body, and filiform; pronotum selliform, extending over the basal articulation of the abdomen, and concealing rudimentary elytra; ovipositor elongated, nearly straight, sword-shaped; tarsi broad, soles concave; third articulation cordate.

SYN.—*Anabrus*, Hald. in Stansb. Expl. Vall. G. Salt Lake, 1852, App. C, 370.

OBS.—The general appearance of the genus *Anabrus* reminds us strongly of *Phalangopsis* proper, from which it is distinguished by its movable and selliform pronotum and the length and shape of the ovipositor. The general proportions of the body and limbs are more elongated than in *Phalangopsis*. The structure of the tarsi, which is not apparent upon a first glance, affords other differences not less important, between *Anabrus* and *Phalangopsis*, when studied comparatively.

2. ANABRUS HALDEMANII, Girard.
ZOOLOGY, Pl. XV, figs. 5-8.

SPEC. CHAR.—Antennæ long and filiform, reaching posteriorly the base of the ovipositor; pronotum short, broad; femora smooth. Yellowish; feet and ovipositor reddish purple. Posterior margin of pronotum black, with two parallel black bands on the posterior third of its length.

DESCRIPTION.—The abdomen above exhibits ten segments or articulations; the anterior or basal one being, as stated above, covered by the posterior prolongation of the pronotum. Beneath there are seven subquadrangular plates, situated opposite to the seven middle upper segments. The posterior segments enclose another piece bearing two spine-like abdominal appendages—one on each side. The ovipositor is as long as the abdomen, and entirely smooth. The base of the antennæ is situated above the eyes, and inserted upon an angular movable piece. The joints composing these organs are very short, and provided with minute setæ. The tibiæ are provided with four rows of spines, two anterior and two posterior; the internal posterior row being the stoutest. The posterior rows are more densely set with spines, whilst the latter are scattered and alternate with each other in the anterior rows. The first and cordate joint of the tarsi is the longest; the second is the shortest; and from the middle of the third, a fourth slender and long joint arises, slightly convex above, and terminating in two spines or claws curved inwards and outwards.

The ground-color above and below is yellowish; the antennæ, limbs, and ovipositor are of a reddish purple. The posterior margin of the pronotum is black. Two parallel black vittæ, enclosing a narrow yellow one, are observed on each side of the dorsal line, upon the posterior third of the pronotum. The posterior portion of the upper abdominal segments is occasionally of a deep-brown hue.

This species differs from *Anabrus simplex*, Hald., by a proportionally much shorter pronotum.

One specimen, caught June 27th, is recorded as "green and white."

Plate XV, fig. 5, represents *Anabrus haldemanii* in a profile view and of the size of life.

Fig. 6 is a front view of the head.

Fig. 7, a tarsus from above.

Fig. 8, a tarsus from below.

III. BRACHYPEPLUS, Charp.

GEN. CHAR.—Body acrydoid; elytra and wings rudimentary; antennæ rather short; pronotum tricarinated; surface between the carinæ granulated. Second joint of tarsi very short; first and third elongated; last one terminating by two curved claws, between which is situated a sub-circular fleshy disk.

SYN.—*Brachypeplus*, Charp. Orth. descr. et pict. Fasc. IX, 1843, Tab. li.

OBS.—This genus, established by Toussaint de Charpentier in his *Orthoptera descripta et picta*, was not characterized, owing, perhaps, to the fact that one species only was known and described by him under the name of *B. virescens*, said to inhabit "Mexico." It may easily be distinguished from the one we shall describe by its much shorter antennæ and slenderer tarsi; also by its color, which is deep-green, with a few brown spots on the pronotum, and a double series of these along the upper part of the abdomen.

3. BRACHYPEPLUS MAGNUS, Girard.
ZOOLOGY, Pl. XV, fig. 1-4.

SPEC. CHAR.—Reddish brown; elytra dotted with black; antennæ bluish brown; femora and tibiæ reddish; tarsi purplish; spines black towards tip; femora sub-fusiform; a carina along the upper and middle region of the abdomen.

DESCRIPTIÓN.—The pronotum is one-third of the length of the abdomen, overlapping posteriorly the anterior abdominal segment entirely and half of the second. The entire number of abdominal segments or articulations is eleven, carinated upon their medial line, and continuing the medial carina of the pronotum all along the middle region of the abdomen above. Antennæ a little longer than the pronotum, and composed of about twenty short joints. The tibiæ are shorter than the femora, and provided, the two anterior pairs internally, and the posterior pair externally, with two rows of spines, the inner row the strongest. The femora are sub-fusiform; the posterior ones a little broader than thick, but never as much compressed as in *B. virescens*, in which these organs present sharp edges. The tarsi are all tetramerous: the first article is the stoutest and the longest, the second being quite short;

the third is more slender, and the fourth the smallest, terminating into two curved spines or claws, between which is a subcircular fleshy disk. The rudimentary elytra are subovoidal, not extending backwards to the posterior margin of the third abdominal segment.

The ground-color, as preserved upon specimens in alcohol, is yellowish brown; black dots and spots are scattered over the rudimentary elytra. The antennæ are bluish brown; the femora and tibiæ reddish, and the tarsi purplish, whilst the spines are black.

This species differs from *B. virescens* by its proportionally longer antennæ, shorter pronotum, and less compressed femora. The general shape of the body is in every respect proportionally longer than in the latter species.

Two specimens were collected on the 7th of July—one "green," the other "reddish brown."

Plate XV, fig. 1, represents *Brachypeplus magnus* in natural size.

Fig. 2, front view of the head.

Fig. 3, a tarsus from above.

Fig. 4, same from below.

We refer to *Brachypeplus virescens* two specimens; one collected on the 12th of June, and which was "green above, white beneath, with yellow and black stripes on the back;" another specimen, a little smaller, caught June 21st, was "green and brown."

ARACHNIDIANS.

BY CHARLES GIRARD.

I. ARANEIDÆ.

1. MYGALE HENTZII, Girard.

ZOOLOGY, Pl. XVI, 1-3.

SPEC. CHAR.—Blackish brown; densely studded with hairs. Cephalothorax subcircular, with a median and transversely elliptical infundibulum upon its posterior half, whence shallow grooves radiate towards the periphery. Abdomen ovoid. Palpi composed of five joints besides the maxillæ, a hook in the male. Legs six-jointed.

DESCRIPTION.—This species is one of the largest of the genus hitherto found within the limits of the United States. The specimen figured, however, is much below the usual size. The cephalothorax is subcircular in shape, a little broader in the male than in the female. The eyes are disposed as in fig. 3, on a little eminence near the anterior margin, and upon the midial line. On the posterior half of the same region, on a line with the eyes, is a transverse infundibulum, sometimes subcrescentic, convex posteriorly. Shallow and sometimes irregular grooves radiate from that centre towards the margin of the cephalothorax. The abdomen is ovoid; considerably larger in the female than in the male. The labrum is quite small. The cheliceræ are robust, regularly arched, terminated by a rather slender hook, similarly curved, and movable upon the cheliceræ. The palpi are six-jointed; the basal joint, functioning as maxilla, is robust, and not otherwise distinguished from the following, except that it is provided along its inner margin with a brush-like series of hairs. The second joint is very short; the third is the longest; the fourth is a little larger than the second; the fifth a little shorter than the third; the fourth shorter than the fifth; the sixth is the size of the second, but differently shaped, being rounded at its extremity, at the inferior surface of which exists a hook, very stout at the base, tapering into an acute point curved downwards and outwards. In the female the sixth joint of the palpi is as long and of the same shape as the fifth, and deprived of the hook. The fourth pair of legs is the longest; the first pair comes next; the second pair is the smallest. They are all six-jointed, the first joint short and robust.

The second joint is the longest; the third the smallest; the fifth is, after the second, the next in length; then the fourth, and finally the sixth. The external pair of fusi, or spinning apparatus, is slender, and, as usual, three-jointed; the internal pair is very small, and not conspicuous. The whole surface of the body and legs, above and below, is densely covered with fine setose hairs. The color is uniform blackish brown.

The *Mygale hentzii* is the large black spider known in the Southwest as the tarantula, where its bite is greatly dreaded.

A female specimen was collected on the 17th of May, on an open, barren prairie between Camps 2 and 3. Other specimens of both sexes were taken on the 28th of June, near the head of south fork of Red river.

Plate XVI, fig. 1 represents *Mygale hentzii* seen from above. Fig. 2 is an underview to exhibit the labrum (l), the maxillæ (m), the cheliceræ (c), and the palpi (p), also to show the fusi (f). Fig. 3 represents the disposition of the ocelli.

2. Lycosa pilosa, Girard.

ZOOLOGY, Pl. XVI, figs. 4 and 5.

SPEC. CHAR.—Hairs of a yellowish brown color, covering the upper parts. Beneath black; cephalothorax subpyriform; abdomen ovoid. Palpi composed of five joints besides the maxillæ; terminal joint provided beneath with two small spines. Legs very long and slender; all six-jointed.

DESCRIPTION.—Of all the American *Lycosa* hitherto described the present species is the one in which the legs are the longest and the most slender. The size of the cephalothorax and abdomen is proportionally smaller, however, than in *L. fatifera*, Hentz.

The cephalothorax is longer than broad, elevated on its middle region, and anteriorly very prominent; subpyriform in its general outline; the narrowest part directed forwards. Its surface, when freed from its fur, exhibits shallow grooves radiating from the centre towards the periphery, pretty much in the same manner as in the *Mygale* just described, although much less conspicuous. There is no central infundibulum, which is replaced here by a minute longitudinal furrow about a tenth of an inch in length. The abdomen is ovoid, and, as usual, larger in the female than in the male.

The cheliceræ are stout, with a very slight downwards inflexion, provided with small protuberances upon the inner margin of its anterior extremity, and terminated by a slender hook curved inwardly. The labrum is comparatively small, whilst the maxillæ are stout. The palpi are slender, and composed of five joints. The first joint is very small, inconspicuous; the second is the longest and the most slender of all; the third is somewhat larger than the first, the fourth larger than the third, and the fifth larger than the fourth, which is swollen and sub-concave beneath, provided with two minute hooks inserted upon two tubercles. In the female the palpi are slenderer than in the male, and the last joint is simple and longer than the third. The legs are long and slender, composed of six joints: the hind pair is the longest; the first pair is the next in length; the third pair is the shortest. The third joint is the smallest in the four pairs; the first joint is the next in length, and the stoutest; the second pair is the longest in the three anterior pair; the fifth comes next, then the fourth and sixth. In the posterior pair the fifth joint is the longest; then the second; then the fourth and sixth. The fusi, four in number, are short, intimately grouped, and composed of a single joint. The whole surface of the body and legs, above and below, is densely covered with short hairs.

The color above is uniform grayish brown. The abdomen, cephalothorax, and first joint of legs beneath, are deep black. The second, third, and fourth joints are of the color of the upper parts upon their middle, and black near their articulations. The fifth and sixth joints are almost entirely black. The extremity of the chelicerae and palpi are black beneath. When the hairy covering is removed, the color is a uniform chestnut-brown.

The color may present some variations; thus in the notes of Captain Marcy, one is described as having "the back brown, belly dirty white, head and legs red."

One specimen preserved in alcohol exhibits a reddish band down the middle of the cephalothorax, and two black vittæ, one on each side of the abdomen. The cephalothorax beneath is reddish; and on the abdomen there are two elliptical light spots.

Specimens were collected the 16th of May on the open prairie, between Camps 1 and 2; and on the 19th of June, on Canadian river, Arkansas.

Plate XVI, fig. 4, represents the trophi, showing the labrum (l), the maxillæ (m), the palpi (p), and chelicerae (c). Fig. 5 exhibits the disposition of the ocelli.

II. TARANTULIDÆ.

THELYPHONUS EXCUBITOR, Girard.

ZOOLOGY, Pl. XVII, fig. 1-4.

SPEC. CHAR.—Blackish brown above, deep chestnut beneath; upper surface of body and legs minutely granular; beneath smooth, with scattered minute imprinted dots. First and second articles of the palpi very granular, remaining ones with a few granules and numerous imprinted dots. Caudal appendage very much developed, and composed of about fifty joints.

DESCRIPTION.—There is a very great resemblance between this species and *T. giganteus*.* The only striking difference which exists between them is to be found in the structure of the palpi and in the length of the caudal appendage.

The cephalothorax is elongated, narrowest anteriorly, where it assumes almost a triangular shape. Its posterior margin is subtruncated, slightly concave in the middle. The central portion of the anterior third of the cephalothorax presents a perfectly plane surface, with a medial furrow, as it were; whilst posteriorly it is depressed, and sloping towards the margins, the surface showing shallow depressions, one upon the middle line, and more regular than the lateral ones. Near the anterior extremity, and in a subcircular depression on each side of a medial, smooth, and rounded elevation, are found the ocelli, circular, large and black. In advance of these ocelli, the rostrum is almost abruptly truncated, as seen in the centre of fig. 3. From the anterior ocelli to the lateral ones extends a linear series of granules, terminating upon the tuberculous elevations, upon which are seen three yellowish ocelli grouped, as exhibited in fig. 2.

The cheliceræ are robust, but very slightly bent, composed of one large joint and a conical, curved, and acute spine; to the inner side of which are attached brushes of quite elongated and reddish setæ. Palpi long and robust, in the shape of arms, and composed of six joints. The first joint is seen only from below (fig. 4, a), and exhibits a subtriangular and flat surface, terminated anteriorly by a conical point. The second joint is smaller than the first, scarcely to be seen viewed from below, but developed upon its upper surface into a flattened and irregular disk,

---

* See *Guérin's* Magazin de Zoologie, 1835, Class VIII, for an illustrated monograph of the genus *Thelyphonus* by H. Lucas.

provided upon its anterior margin with five conical spines, varying in size: seen in front (fig. 3, b), it is elevated almost vertically from the horizontal position of the first. The third joint is the longest of all, slightly curved, and provided inwardly with two minute spines—one above, the other below. The fourth joint is somewhat shorter than the third, but is much longer than broad, subcylindrical, slightly bent, and provided at its inner, anterior, and upper edge, with a prominent, conical, and straight spine. The fifth joint is of the length of the fourth, but slenderer, and provided anteriorly with a stout and shorter spine. Finally, the sixth joint is a subconical and spiny processus, moving against the spine of the fifth joint, constituting a forceps, and used as such to seize prey. The thoracic appendages (feet) are long and slender, especially the anterior and posterior pairs. The anterior pair may be readily distinguished from the three others, in not being provided with hooks upon their extremity. Its function is rather that of a pair of palpi than that of ambulatory organs. The first and second joints are short and stout; the third, fourth, and fifth long and slender; the fourth and fifth almost equal in length, and longer than the third. Eight small joints, together equal in length to the third, terminate these appendages. The three others are constructed upon the same plan, all having nine joints and terminal hooks, generally two in number. The first, second, and third joints are similar to those of the anterior pair; the third, however, is the longest; the fourth is but a little longer though slenderer than the second, and slightly curved; the fifth is much slenderer and a little shorter than the third. Next come four small joints, together smaller in length than the fifth, and provided upon their anterior margin with minute spines. The second of these four, or the seventh in the series, is the longest of the four; the third is the smallest; the first and fourth are equal in length, the latter much slenderer. Two hook-like and slender spines terminate these organs.

The abdomen is longer than the cephalothorax, oval in shape, though depressed, and composed of eight very distinct segments and a half, the anterior one. The stigmatiform bodies are quite conspicuous above (seven pairs), and below (four pairs). The anterior half segment is not seen from below. The seventh segment exhibits laterally a second pair of stigmatiform bodies, less conspicuous, however, than the others. The posterior segment has also faint traces of an analogous pair. The two first caudal rings are very narrow; the third is as large as the two others together. The filiform appendage is very long, and composed of about fifty joints.

The upper surface of the cephalothorax and abdomen is covered with minute granules extending over the palpi, being particularly dense on the three first articles, and over the three first joints of the thoracic appendages also. Minute impunctures are seen upon the remaining articles and joints, and also scattered upon the inferior surface of the appendages and body. Minute setæ are scattered over the appendages of the cephalothorax and abdomen, more densely towards their extremities.

The color is uniform blackish brown above, and deep chestnut beneath.

One specimen of this animal was collected.

Plate XVII, fig. 1, represents, seen from above, *Thelyphonus excubitor* the size of life.

Fig. 2 gives the position, number, and relative size of the ocelli.

Fig. 3 is a front view, exhibiting in the centre the cheliceræ and the three first articles (a, b, c) of the palpi.

Fig. 4 represents the anterior portion of the cephalothorax from below: *a*, first article, *b*, second article, and *c*, third article of the palpi; and *d*, anterior pair of feet.

III. SCORPIONIDÆ.

Although the collections made in the valley of Red river contained no specimens of this group of arachnides, we have brought them here to notice, satisfied as we are that they exist in that locality.

Scorpions are found in the southern Atlantic States, all along the Gulf of Mexico, through Texas and New Mexico to California, and through Louisiana to Arkansas.

1. Scorpio (Telegonus) boreus, Girard.

Zoology, Pl. XVII, figs. 5–7.

Spec. char.—Body greenish yellow; thoracic and caudal appendages yellowish. Lateral ocelli in close contiguity; posterior one the smallest. Median ocelli situated on the sides of an elongated and black elevation. Chelicerae terminated by a serrated claw. Palpi robust, shorter than the body. Caudal appendage as long as the body, the spine excepted. Abdominal comb with eighteen laminæ.

Description.—The general form of the body is fusiform, anteriorly and posteriorly tapering. The cephalothorax proper is subquadrilateral, longer than broad, narrower anteriorly than posteriorly; both of these extremities linear; lateral margin somewhat undulated. Its surface is carved with a few undulating grooves, giving to the rest an undulated appearance; and over the whole, minute granules. The median ocelli are black, situated a little in advance of the middle of the length of the thorax, and placed on the sides of an elongated, little, and black eminence, divided longitudinally by a groove. The lateral ocelli are set close together and situated near the anterior margin of the cephalothorax; the posterior one is much the smallest: they are represented with their relative proportions in fig. 7. The cheliceræ are stout, two-jointed; the second being the largest, and is terminated by a minutely serrated claw. The palpi are five-jointed; the first joint is short and stout, and fulfils the function of jaws without denticulation. The second is the smallest. The third and fourth are more elongated; the third a little longer than the fourth. They are angular, the angles being margined with dense rows of minute granules. The fifth joint or hand (carpus) is stout and swollen, exhibiting eight undulating ribs (four above and four below), upon which is a row of minute granules. Two rows above and below are seen extending along the spiny immovable processus of the hand, constituting, with a movable spine, a slender chela or claw, slightly curved inwards. Scattered setæ may be seen on the whole length of the palpi; and also on the thoracic appendages (feet). The latter are slender; the fourth pair is the longest; the first pair the smallest, the second and third pairs being of intermediate proportions; the second longer than the first, and the third longer than the second. They are all flattened, seven-jointed, and terminated by minute hooks. The third joint is in every one the longest and most slender; the fifth, sixth, and seventh are small, the seventh being the smallest of all. There are generally three terminal hooks; occasionally minute spines may be seen near the articulation of the sixth and seventh joints. The first joint is the stoutest, and in the first pair of these appendages it has something to do with mastication, functioning perhaps as a lower lip. The abdominal combs are slender and elongated, and composed of a transverse triarticulated piece, and of eighteen little laminæ attached to it. The dorso-abdominal shields, seven in number, increase in size from forwards backwards, the anterior one being the narrowest of all. Their surface exhibits minute granules not very conspicuous. There are only five ventral shields, nearly equal in size; the posterior one somewhat different in shape, and not provided with stigmata. The caudal appendage (tail) is as long as the body, and composed of five joints and a poison bag. The two first joints are the smallest, the fifth being the longest; the poison bag is swollen up and provided with a slightly curved and acute hollow spine. The upper part of each joint is concave or grooved, whilst the inferior part is convex. They are carinated, and rows of conspicuous granules are observed along the carinæ.

The color of the body above is uniform greenish yellow; the thoracic appendages (feet) are yellowish, whilst the palpi and caudal appendage (tail) reflect a reddish shade upon the yellow ground.

The specimen figured was collected in the Valley of the Great Salt Lake of Utah, by Capt. Howard Stansbury.

A much smaller specimen was brought from Eagle Pass, Texas, by Mr. Arthur Schott, of the United States and Mexican boundary.

Plate XVI, fig. 5, represents, size of life, *S.* (*Telegonus*) *boreus* seen from above.

Fig. 6 is a view from beneath, to show the abdominal combs, first abdominal segment, and origin of fourth and third pairs of feet.

Fig. 7 represents the distribution of the ocelli.

2. Scorpio (Atreus) californicus, Girard.

General form of body and appendages slender when compared to the preceding species. The tail is almost twice the length of the body; there is not the same disproportion of length between the first and second joints and the remaining ones. The carinæ and rows of granules are much less conspicuous. The cephalothorax and dorso-abdominal shields exhibit carinæ and rows of granules not only on the palpi, but likewise on the feet. Rows of granules may be seen along the angular projections or carinæ. The chelæ are much slenderer, the hand (carpus) and poison bag much smaller. An exceedingly minute spine may be observed on the poison bag under the sting. The lateral ocelli are situated more anteriorly, more apart from each other, and equal amongst themselves. The abdominal combs are composed of twenty laminæ.

Color light brown; palpi and tail deeper; upper part of abdomen blackish, with a median light vitta.

One specimen was collected in California and presented by Dr. Stone to the Smithsonian Institution.

3. Scorpio (Atreus) sayi, Girard.

Syn.—*Buthus vittatus*, Say, Jour. Acad. Nat. Sc. Philad. II, 1821, 61.

Upon a close examination of several specimens of this species obtained from western Florida, we satisfied ourselves that it belongs to the subgenus *Atreus* instead of *Buthus*, in which it was placed by Thomas Say. It so happens that the specific name of *vittatus* has since been given by Guérin to another South American species of scorpions; and if we propose here to replace Say's specific name, against the received law of priority, we would remark that when full grown, the vittæ entirely disappear, and the color becomes uniform deep reddish brown, the legs and under surface being lighter. In this species the tail is once and a half the length of the body. The palpi are proportionally small, and in the young, exiguous. The chelæ are slender, slightly curved, with an undulation at their base, but without marked denticulations. The upper surface is finely granular. There are from thirty to thirty-two laminæ to the abdominal combs. "Fuscous, with three fulvous vittæ, sides black," applies strictly to the immature state.

Specimens of this species were sent from Pensacola, Florida, to the Smithsonian Institution, by Dr. Jeffrey, U. S. N., and Dr. J. F. Hammond, U. S. A.

A species very closely allied, if not identical with *Scorpio (Atreus) sayi*, is not uncommon in Texas, where several specimens were collected by Lieut. D. N. Couch, U. S. A.

IV. PSEUDOSCORPIONIDÆ.

Observations upon Galeodes subulata of Thomas Say.

Two species of this genus are described by the same author in Major Long's Expedition;* one under the name of *Galeodes pallipes*, the other under that of *G. subulata*, the only difference between them consisting in the structure of the cheliceræ, which in *G. pallipes* are terminated by arcuated claws, armed within with many robust teeth, whilst in *G. subulata* the upper claw is nearly rectilinear, and the lower one alone possessed with two robust teeth.

Having but one individual of this genus at our command, we are not prepared to decide upon the question of the validity of both species. The specimen before us answers to Say's characters of *G. subulata;* and being perfectly satisfied that it belongs to the latter species, we propose to describe it a little more at length than was done by its discoverer.

---

*Account of an expedition from Pittsburg to the Rocky Mountains, performed in the years 1819 and '20. Vol. II, 1823, p. 3.

The entire length, from the tip of the cheliceræ to the end of the abdomen, is one inch and a quartér, the abdomen itself forming about one-half of that length. The cephalothorax is composed of three distinct segments; the anterior one much the largest, giving points of attachment to the parts of the mouth, to the palpi, and the two anterior pairs of legs; to the second thoracic segment is attached the third pair of legs, and to the third segment the fourth pair. The anterior segment of cephalothorax, seen from above, is subrhomboidal and smooth. At its anterior margin are situated the two ocelli, separated from each other by a deep groove. The cheliceræ are very stout, and composed of one single joint densely covered with setose hairs, and terminated each by two spines, one above (finger of some authors), rigid, and another below (the thumb), moving vertically against the upper. The latter is compressed, acute, almost rectilinear, and smooth; the inferior one is subconical, curved upwards, acute towards the point, and provided at its base inwardly with two spiny small processes. The palpi are proportionally robust, stouter and longer than the three anterior pairs of legs; somewhat shorter than the fourth pair, but of a stouter appearance, as all the joints, four in number (the maxillæ excepted), preserve the same diameter. They are covered on their whole length with hairs similar to those on the cheliceræ. The maxillæ are subtriangular, provided only with brushes of hairs. The next joint (the joint of the palpi) is very small and triangular; the second is the longest; the third is the next in length; then the fourth, the tip of which exhibits a minute smooth tuberculiform knob. The first pair of legs is the most slender of the thoracic appendages, and about the length of the third pair; the basal joint is quite short; the second is the shortest of all; the third is the longest; the fourth, fifth, and sixth smaller in the order enumerated. The last joint terminates like the palpi, bluntly. This anterior pair of legs is called by some *second pair of palpi*, upon the ground that their structure is most alike. The three remaining pairs of thoracic appendages are seven-jointed, thus composed of one joint more than in the first pair and palpi, and furthermore terminated by two minute curved claws. The first, second, and third joints are short, stoutish, and subequal; the remaining are longer and slenderer, the fourth being the longest, and the other diminishing gradually. They are covered upon their whole length with hairs similar to those which cover the palpi, but perhaps less densely so. The abdomen is subovoid, being a little depressed; it is densely hairy above and below, and composed, as usual, of nine segments or annuli.

Collected on June the 10th.

MYRIAPODS.

BY CHARLES GIRARD.

---

1. SCOLOPENDRA HEROS, Girard.

ZOOLOGY, Pl. XVIII.

SPEC. CHAR.—Twenty-one pairs of grallatory appendages, composed of five segments or articulations, and a conical terminal spine, more or less curved. Back bicarinated; beneath, flat and grooved. Antennæ composed of twenty-five joints; color uniform dark-reddish brown; lighter beneath.

DESCRIPTION.—The general form of the body is depressed, subconcave above, flat beneath. It is composed of twenty-one annuli, segments or rings, each of which bears one pair of locomotory appendages, (feet). The middle region of the back presents a slight double carina and last segment. The intermediate area is rather flattened, whilst each running parallel the whole length of the body, very faint on the first side is gently sloping towards the exterior margin. At the inferior surface, two longitudinal furrows or grooves may be seen extending the whole length of the body, and dividing the abdominal disk into three almost equal parts. The stigmata are transversely elongated, and situated immediately beneath the lateral margin of the dorsal shields of each segment. The insertion of the locomotory appendages takes place immediately above the lateral margin of the abdominal shields of each segment. The locomotory appendages are as numerous as the segments of the body—twenty-one pairs constructed alike; that is to say, composed of five joints and a curved terminal spine. A minute spine may occasionally be seen at the anterior margin of the fourth and fifth joints. The third and fourth joints are longer than the first and second; the fifth is always the smallest: these organs are tapering rapidly towards their extremity. In the caudal pair, the first and second articles or joints are longer than the third and fourth; the first one is, moreover, provided with a spiny process along its inner margin. Its general shape and directing distinguishes it, likewise, from all the other pairs.

The second segment is quite short, the shortest of all, and contrasts strangely with the others, which preserve regular proportions, gradually diminishing from the middle of the length towards both extremities, with but few exceptions. The first segment or ring is one of these, being the shortest after the second; its anterior margin is subcrescentic, the concavity of which receives the cephalic shield or disk (head). Besides the anterior pair of locomotory appendages, it gives a point of attachment to a pair of robust and two-jointed forceps, functioning as a pair of jaws for seizing and holding the prey. The central piece is large and subtriangular, the anterior margin of which is denticulated, (the second lip of some authors). That second or external lip (labrum) is formed by the union of two pieces, which are separate in the young, where they constitute a third joint to the forceps-jaws, the second lip then being also separate, and existing as a limina already denticulated anteriorly. The next joint is short and stout; the second is a conical and tapering spine, curved inwardly and perforated, as it is well known for the passage of a venomous fluid, not otherwise dangerous.

The cephalic disk itself, seen from above, is subcircular in shape, projecting slightly between the antennæ, and showing upon its surface traces of the dorsal carinæ alluded to above. To its inferior surface we find attached two pairs of mandibles and one pair of palpi. In proceeding from outwards inwards, we will find immediately behind the forceps jaws the palpi (little feet, sometimes called), composed of four flattened joints and a minute, curved, and terminal hook. They are united at their base by the means of two additional central pieces. The second joint is the longest, and slightly bent. The exterior pair of mandibles, the one next to the palpi, is composed of four joints, the first being almost as long as the three remaining ones; the fourth is rounded, presenting an inner concave surface with a sharp terminal margin. They are united upon their middle by a lanceolated ligula. The inner pair of mandibles is composed of two pieces; the first irregularly shaped, the second subcircular concave, subcircular and margined anteriorly by small spines, four or five in number, constituting a denticulated margin.

In the anterior margin of the cephalic disk are inserted the antennæ, composed of twenty-five joints gradually diminishing in thickness, and increasing in length towards the extremity, which is filiform. Exteriorly to the antennæ, and close to the margin of the disk, are situated the ocelli, four on each side, as usual in the genus, and disposed as represented in figure 5.

The inferior surface of the last ring differs from the others in having a much smaller shield, and in being provided on each side with a stout, subconical spine, directed backwards.

An immature specimen, one-third of the length of the one figured, has the same number of segments or annuli, the same number of feet, and the same general structure.

One individual of this species was collected, on the 15th of June at Sweet-water creek; others were found in July, between the south fork of Red river and Otter creek.

Plate XVIII, fig. 1, represents *Scolopendra heros* size of life, seen from above.

Fig. 2. The head from below.
Fig. 3. Posterior extremity from below.
Fig. 4. A medial segment to show the attachment of feet.
Fig. 5. Disposition of ocelli on left side.

2. Julus ornatus, Girard.

Spec. char.—Ground-color bluish black; segments narrowly margined posteriorly with reddish; anterior margin of segments rather blue, whilst the middle is rather black, thus giving the appearance of three rings of color. The anterior portion, which is covered by the articulation, is fulvous. Feet deep chestnut-brown. Antennæ rufous at base, blackish at tip. Stigmata not conspicuous; marked by a series of small, obsolete blackish spots.

Remarks.—This species is allied to *Julus marginatus* of Say, but its body is proportionally much stouter. The ocelli are disposed upon a subtriangular space quite different in shape. The antennæ themselves are slenderer in proportions. The labrum (upper lip) is also less emarginated than in *Julus marginatus*, and the marginal punctures much less conspicuous.

One specimen was collected, on the 27th of June.

3. Julus atratus, Girard.

Spec. char.—Body, feet, and antennæ, uniform deep blackish brown; antennæ and feet occasionally reddish, as also the labrum and anterior margin of first segment. Posterior third of each segment of a shining black. Stigmata and lateral striæ beneath quite conspicuous.

REMARKS.—Resembles more *Julus ornatus* than *Julus marginatus* in the general proportions of the body, but in the structure of the antennæ and labrum comes nearer to *Julus marginatus*.

Specimens of this species were collected at Prairie Mer Rouge, Louisiana, by James Fairie, Esq., and sent to the Smithsonian Institution.

APPENDIX G.

BOTANY

DESCRIPTION OF THE PLANTS COLLECTED DURING THE EXPEDITION: BY DR. JOHN TORREY.

APPENDIX G.

BOTANY.

BY JOHN TORREY, M. D.

No. 96, St. Mark's Place, New York,
August, 10, 1853.

DEAR SIR: I have examined the collection of plants that you brought from the headwaters of the Red river, towards the Rocky mountains. The flora of this region greatly resembles that of the upper portion of the Canadian. It is remarkable that there occur among your plants several species that were first discovered by Dr. James, in Long's Expedition, and have not been found since until now. Your collection is an interesting addition to the geography of North American plants, and serves to mark more clearly the range of many western species. For particular remarks on the rarer plants, and descriptions of the new species, I refer you to the accompanying list.

At your request I have had some of the rarer plants drawn and engraved, to illustrate your report to Congress.

I am, dear sir,
Yours truly,
JOHN TORREY.

Captain R. B. MARCY.

RANUNCULACEÆ.

CLEMATIS PITCHERI, Torr. and Gr., Fl. 1, p. 10. Witchita Mountains; fl. and fr. July 17.

ANEMONE CAROLINIANA, Walt.; Torr. and Gr., Fl. 1, p. 12. Sources of the Trinity River; May 3.

DELPHINIUM AZUREUM, Michx.; Torr. and Gr., Fl. 1, p. 32. Main Fork of the Red River; fl. May 8—June 16.

PAPAVERACEÆ.

ARGEMONE MEXICANA, Linn.; Torr. and Gr., Fl. 1, p. 61. Common on the upper waters of the Red River; May—June 16.

CRUCIFERÆ.

VESICARIA ANGUSTIFOLIA, Nutt., in Torr. and Gr., Fl. 1, p. 101; Gray, Pl. Lindh. 2, p. 145. Sources of the Trinity River; fl. and fr. May 3.

V. STENOPHYLLA, Gray, Pl. Lindh. 2, p. 149; and Pl. Wright. 1, p. 10, and 2, p. 13. North Fork of the Red River; fr. June 14.

DITHYRÆA WISLIZENI, Engelm., in Wisliz. N. Mex., p. 95; Gray, Pl. Wright. 1, p. 10, and 2, p. 14. Abundant on the headwaters of the Red River; June 23—July 14.

The specimens of this plant collected by Captain Marcy vary considerably in the leaves, which are often nearly entire. The flowers also vary in size; the petals being sometimes nearly one-third of an inch in length. The silicles are larger than in specimens collected in New Mexico by Mr. Wright and Dr. Edwards. They are by no means always deeply emarginate at the base, and sometimes they are slightly notched at the summit.

STREPTANTHUS HYACINTHOIDES, Hook., in Bot. Mag., t. 3516; Torr. and Gr., Fl. 1, p. 78; Gray, Gen. Ill., t. 61. Witchita Mountains to the boundary of the Choctaw Nation; fl. May 31—June 4.

CAPPARIDACEÆ.

POLANISIA GRAVEOLENS, Raf.; Torr. and Gr., Fl. 1, p. 123, and Suppl., p. 669. Witchita Mountains; fl. and fr. July 16. The pods are on a short stipe, and the seeds are more or less rough.

CARYOPHYLLACEÆ.

SILENE ANTIRRHINA, Linn., Torr. and Gr., Fl. 1, p. 191. On the Main Fork of the Red River; fl. May 8.

PARONYCHIA JAMESII, Torr. and Gr., Fl. 1, p. 170; Gray, Pl. Fendl., p. 14. Middle Fork of Red River; fl. May 22.

PORTULACACEÆ.

TALINUM TERETIFOLIUM, Pursh, Fl. 2, p. 365; Gray, Gen. Ill., t. 98. Middle Fork of Red River; fl. May 22, fr. July 5.

MALVACEÆ.

MALVASTRUM COCCINEUM, Gray, Gen. Ill., t. 121; Pl. Fendl., p. 24. *Malva coccinea*, Nutt. *Sida coccinea*, DC.; Torr. and Gr., Fl. 1, p. 235. North Fork of Red River, &c.

CALLIRRHŒ INVOLUCRATA, Gray, Pl. Fendl., p. 15, and Gen. Ill., t. 117. *Malva involucrata*, Torr. and Gr., Fl., p. 226. Middle Fork of Red River; fl. May 22.

C DIGITATA, Nutt. in Jour. Acad. Phil. 2, p. 181; Gray, l. c. Fort Belknap.

LINACEÆ.

LINUM BERLANDIERI, Hook. Bot. Mag., t. 3480; Engelm. in Gray, Pl. Wright. 2, p. 25. Cache creek, and Cross-Timbers of the Red River; May.

L. BOOTTII, Planch., in Lond. Jour. Bot. 7, p. 475; Engelm. l. c. Witchita Mountains; fl. and fr. July 17.

OXALIDACEÆ.

OXALIS VIOLACEA, Linn.; Torr. and Gr., Fl. 1, p. 211. Headwaters of the Trinity River; April 25.

O. STRICTA, Linn.; Torr. and Gr., Fl. l. c. With the preceding.

GERANIACEÆ.

GERANIUM CAROLINIANUM, Linn.; Torr. and Gr., Fl. 1, p. 207. Headwaters of the Trinity, and on Cache Creek; April—May.

ZANTHOXYLACEÆ.

PTELEA TRIFOLIATA, Linn.; Torr. and Gr., Fl. 1, p. 215; *β. mollis*. Torr. and Gr., Fl. 1, Suppl., p. 680. Common on the headwaters of the Red River; fr. June 16.

ANACARDIACEÆ.

RHUS TRILOBATA, Nutt., in Torr. and Gr., Fl. 1, p. 218; Gray Pl. Fendl., p. 28. On the Middle and North Forks of the Red River; in fruit June 1–16.

R. TOXICODENDRON, Linn.; Torr. and Gr., l. c. With the preceding in fruit only.

VITACEÆ.

VITIS RUPERTRIS, Scheele, in Linnæa, 21, p. 591; Gray, Pl. Lindh., 2, p. 165. Witchita Mountains; abundant. The fruit was immature, but had attained nearly its full size in the middle of July. They are said to be ripe in August, when they are about the size of large peas, of a deep purple color, and agreeable to the taste. This species much resembles the summer grape of the Atlantic States.

SAPINDACEÆ.

SAPINDUS MARGINATUS, Willd.; Torr. and Gray, Fl. 1, 255; Gray, Gen. Ill., 2, t. 180. Main Fork of Red River.

This is generally known in Texas and Arkansas by the name of *Wild China*. It is a tree, and attains the height of 20 feet, with a trunk 10 inches in diameter. The wood is of a yellow color.

POLYGALACEÆ.

POLYGALA ALBA, Nutt. Gen. 2, p. 87; Gray, Pl. Wright. 1, p. 38. *P. Beyrichii*, Torr. and Gr., Fl. 1, p. 670. On Suydam Creek, North Fork of Red River; fl. June 6.

P. INCARNATA, Linn.; Torr. and Gr., 1, p. 129. Tributaries of the Washita River; fl. and fr. July 23. This species has not hitherto been found so far west.

KRAMERIACEÆ.

KRAMERIA LANCEOLATA, Torr., in Ann. Lyc. N. York, 2, p. 168; Gray, Gen. Ill., 2, t. 185. Headwaters of the Trinity, and on the Middle Fork of the Red River; fl. May 4–22.

LEGUMINOSÆ.

VICIA MICRANTHA, Nutt., in Torr. and Gr., Fl. 1, p. 271. Cache Creek and Middle Fork of Red River; fl. and fr. May 16–22.

RHYNCHOSIA TOMENTOSA, var. *volubilis*, Torr. and Gr., Fl. 1, p. 285. Tributaries of the Washita River; fl. July 26.

TEPHROSIA VIRGINIANA, Pers.; Torr. and Gr., Fl. 1, p. 295. Witchita Mountains and upper waters of Red River; fl. June 4, fr. July 23.

GLYCYRRHIZA LEPIDOTA, Nutt., Gen. 2, p. 106; Torr. and Gr., Fl. 1, p. 298. Main and North Forks of the Red River; fl. June 6, fr. June 26.

INDIGOFERA LEPTOSEPALA, Nutt., in Torr. and Gr., Fl. 1, p. 298. With the preceding; fl. May 26–June 6.

PSORALEA ESCULENTA, Pursh, Fl. 2, p. 475, t. 22. Mouth of Cache Creek, and Witchita Mountains; May.

P. ARGOPHYLLA, Pursh, Fl. 2, p. 475; Hook. Fl. Bor.—Am. 2, p. 136, t. 53. North and Middle Forks of Red River; fl. May 26–31.

P. FLORIBUNDA, Nutt., in Torr. and Gr., Fl. 1, p. 300. Sources of the Red River; fl. June 2–9.

PETALOSTEMON VIOLACEUM, Michx., Fl. 2, p. 50, t. 37, f. 2; Torr and Gr., Fl. 1, p. 310. With the preceding; June 2–7.

PETALOSTEMON GRACILE, Nutt. in Jour. Acad. Phil. 7, p. 92; Torr and Gr., Fl. 1, p. 309. Cache Creek; May 18.

P. MULTIFLORUM, Nutt., l. c.; Torr. and Gr., l. c. On the Witchit Mountains; fl. and fr. July 15.

PETALOSTEMON VILLOSUM, Nutt., Gen. 2, p. 85; Torr. and Gr., Fl. 1, p. 310. Cache Creek; June 14; flowers not yet expanded.

DALEA AUREA, Nutt., Gen. 2, p. 101; Torr. and Gr., Fl. 1, p. 308; Gray, Pl. Wright. 2, p. 41. Main Fork of Red River; fl. July 5.

D. LANATA, Spreng, Syst. 3, p. 327. *D. lanuginosa*, Nutt., in Torr. and Gr., Fl. 1, p. 307. Big Witchita and on the Main Fork of the Red River; fl. June 27.

D. LAXIFLORA, Pursh, Fl. 2, p. 741; Nutt., Gen. 2, p. 101; Torr. and Gr., Fl. 1, p. 307. *D. pencillata*, Moricand, Pl. Nouv. Amer., t. 45. Common on all the upper waters of the Red River; May–July.

AMORPHA CANESCENS, Nutt., Gen. 2, p. 92; Torr. and Gr., Fl. 1, p. 306. Witchita Mountains; fl. May 30.

ASTRAGALUS NUTTALLIANUS, DC. Prodr. 2, p. 289; Torr. and Gr. 2, p. 234. Upper waters of the Red River; fl. and fr. May 5. The flowers are larger than usual in this species.

A. CARYOCARPUS, Ker. Bot. Reg., t. 176; Torr. and Gr., Fl. 1, p. 331. Headwaters of the Trinity. May 2; in flower only.

OXYTROPIS LAMBERTI, Pursh, Fl. 2, p. 740; Torr. and Gr., Fl. 1, p. 339. With the preceding; fl. in May.

DESMODIUM SESSILIFOLIUM, Torr and Gr. 1, p. 363. Witchita Mountains. The specimens of this plant collected by Captain Marcy are in a state of remarkable *fasciation*. The branches of the panicle are coalesced, (sometimes almost to the summit,) into a broad flat mass, which is covered with sessile flowers and fruit.

CLITORIA MARIANA, Linn.; Torr. and Gr., Fl. 1, p. 290; Torr., Fl. N. York, 1, p. 163, t. 24. On the Washita; fl. July 27.

BAPTISIA AUSTRALIS, R. Br.; Torr. and Gr., Fl. 1, p. 385. Sources of the Red River; fl. and fr. June 6–10.

B. LEUCOPHÆA, Nutt., Gen. 1, p. 282; Torr. and Gr., l. c. Common on the upper tributaries of the Red River; fl. April, fr. May.

HOFFMANSEGGIA JAMESII, Torr. and Gr., Fl. 1, p. 293; Gray, Pl. Lindh. 2, p. 178. With the preceding; fl. and fr. June 14–24.

CASSIA CHAMÆCRISTA, Linn.; Torr. and Gr., Fl. 1, p. 395. Tributaries of the Washita; fl. July 22.

SCHRANKIA UNCINATA, Willd.; Torr. and Gr., Fl. 1, p. 400. Mouth of Medicine River, &c.; fl. April.

ACACIA LUTEA, Leavenw.; Torr. and Gr., Fl. 1, p. 403. On the Witchita Mountains; fl. and fr. July 14. The leaves are remarkably sensitive.

ROSACEÆ.

SANGUISORBA ANNUA, Nutt., in Torr. and Gr., Fl. 1, p. 429. *Poterium annuum*, Hook. Fl. Bor.—Am. 1, p. 198.

ONAGRACEÆ.

ŒNOTHERA RHOMBIPETALA., Nutt., in Torr. and Gr., Fl. 1, p. 493; Kunze, in Linnæa, 20, p. 57. Main Fork of Red River; fl. June 24.

Œ. SINUATA, Linn.; Torr. and Gr., Fl. 1, p. 294. Witchita Mountains and upper tributaries of the Red River; May–June.

Œ. SPECIOSA, Nutt., in Jour. Acad. Phil. 2, p. 119; Torr. and Gr., Fl. l. c. Big Witchita; fl. May 8. Middle Fork of the Red River; fr. June 21.

Œ. LAVANDULÆFOLIA, Torr. and Gr., Fl. 1, p. 501; Hook. Lond. Jour. Bot. 6, p. 223, Gray, Pl. Wright. 1, p. 72. Big Witchita and North Fork of Red River; fl. May 8, fr. June 6. The leaves in all our specimens of this rare species are nearly glabrous, about one inch and a half long, and 2–3 lines wide, with the apex rather acute. The fruit is well described by Hooker, (l. c.)

Œ. SERRULATA, Nutt. Gen. 1, p. 246; Torr. and Gr., Fl. 1, p. 501. Common on the upper tributaries of the Red River; May–June.

GAURA COCCINEA, Nutt. Gen. 1, p. 249; Torr. and Gr., Fl. 1, p. 518. North Fork of Red River; fl. June 6.

G. VILLOSA, Torr. Ann. Lyc. N. York, 2, p. 200; Torr. and Gr., Fl. 1, p. 518; Gray, Pl. Wright, 1, p. 73. Witchita Mountains; fr. July 14. The ripe fruit is not always reflexed. It is (including the stipe) about

7 lines long, ovate, strongly tetraquetrous, abruptly contracted at the base, and 2–4-seeded: the seeds more or less imbricated.

LOASACEÆ.

MENTZELIA NUDA, Torr. and Gr., Fl. 1 p. 535; Gray, Pl. Fendl., p. 47, and Pl. Wright. 1, p. 73; *Bartonia nuda,* Nutt. Gen. 1, p. 297. Witchita Mountains; fl. June 22.

CUCURBITACEÆ.

CUCURBITA PERENNIS, Gray, Pl. Lindh. 2, p. 193; and Wright. Pl. 2, p. 60. *C. fœtidissima,* H. B. and Kunth? *Cucumis perennis,* James, in Long's Exped. 2, p. 20; Torr. and Gr., Fl. 1, p. 543. North fork of the Platte; fl. June 6. Although the cultivated plant seems to be dioecious not unpleasant to the smell, Mr. Wright says, (*vide* Gray, l. c.) that in a wild state it is " certainly monoccious, and exhales an unpleasant smell when bruised;" so that it does not differ from the description of *C. fœtidissima,* except that the latter is said by Kunth to be an annual, which may be a mistake. The flowers are as large as those of the common pumpkin.

SICYDIUM, sp. nov.? Fruit $1\frac{1}{2}$ inch in diameter, globose, sessile. Seeds $\frac{1}{8}$ larger than in *S. Lindheimeri,* and more turgid. On the Main Fork of Red River; fr. July 11.

GROSSULACEÆ.

RIBES AUREUM, Pursh, Fl. 1, p. 164; Torr. and Gr., Fl. 1, p. 552. North Fork of Red River; fr. June 4.

UMBELLIFERÆ.

ERYNGIUM DIFFUSUM, Torr., in Ann. Lyc. N. York, 2, p. 207; Torr. and Gr., Fl. 1, p. 603. Witchita Mountains; fl. June 14. This rare species has not been found before, since it was first discovered by Dr. James, more than thirty years ago. It is rather doubtful whether it is diffuse, except, perhaps, when it is old. The specimens of Captain Marcy are less branched than the original one from which the description in the Flora of North America was drawn.

LEPTOCAULIS ECHINATUS, Nutt., in DC. Prodr. 4, p. 107; Torr. and Gr., Fl. 1, p. 609. Headwaters of the Trinity; April 2.

POLYTÆNIA NUTTALLII, DC. Umb., p. 53, t. 13, and Prodr. 4, p. 196; Torr. and Gr., Fl. 1, p. 533. Middle Fork of Red River; fl. June 1. Witchita Mountains; fr. July 16.

EURYTÆNIA TEXANA, Torr. and Gr., Fl. 1, p. 633. Main Fork of Red River; fr. June 11. This plant has hitherto been found only by the late Mr. Drummond, who discovered it in Texas more than twenty years ago. It is an annual, about two feet high; the fine striæ of the stem and branches are roughened upward, with minute points. The umbels are compound and spreading. Flowers minute. Petals white, broadly orbicular, waved on the margin, deeply emarginate, with an inflexed point. Fruit about one-third larger than in Drummond's Texan specimen.

RUBIACEÆ.

OLDENLANDIA ANGUSTIFOLIA, Gray, Pl. Wright. 2, p. 68. *Houstonia angustifolia*, Mich. Fl. 1, p. 85; *Hedyotis stenophylla*, Torr. and Gr., Fl. 2, p. 41. Tributaries of the Main Fork of Red River; fl. May—June.

VALERIANACEÆ.

FEDIA RADIATA, β. LEIOCARPA, Torr. and Gr., Fl. 2, p. 52. Upper Red River.

COMPOSITÆ.

LIATRIS SQUARROSA, Willd.; Torr. and Gr., Fl. 2, p. 68; Sweet Fl. Gard., t. 44. Tributaries of the Washita River; fl. July 22–24.

L. ACIDOTA, Engelm. and Gray, Pl. Lindh., p. 10; Gray Pl. Wright. 1, p. 83. *L. mucronata*, Torr. and Gr., Fl. 2, p. 70; not of D. C. On the Washita; July 27.

SOLIDAGO ODORA, Nutt.; Torr. and Gr., Fl. 2, p. 219. Witchita Mountains; July 16.

S. MISSOURIENSIS, Nutt. in Jour. Acad. Philad. 7, p. 32, and Trans. Amer. Phil. Soc. (n. ser.) 7, p. 327; Torr. and Gr., Fl. 2, p. 222. With the preceding.

ARTEMISIA FILIFOLIA, Torr. in Ann. Lyc. N. York, 2, p. 211; Torr. and Gr., Fl. 2, p. 417. Upper tributaries of the Red River; May. An abundant shrub, of a grayish white aspect, with numerous branches, and crowded, slender leaves. This is one of the numerous species called *sage* by the hunters. It is found from the plains of the Upper Missouri to the Valley of the Rio Grande, and west to the Colorado.

ACHILLEA MILLEFOLIUM, Linn.; Torr. and Gr., Fl. 2, p. 409. With the preceding. It is the woolly form that almost exclusively occurs west of the Mississippi.

ZINNIA GRANDIFLORA, Nutt. in Trans. Amer. Phil. Soc. (n. ser.) 7, p. 348; Torr. and Gr., Fl. 2, p. 298; Torr. in Emory's Rep., t. 4, Gray, Pl. Fendl., p. 81. Main Fork of Red River; fl. July 2.

RIDDELLIA TAGETINA, Nutt. l. c., p. 371; Torr. and Gr., Fl. 2, p. 362; Torr. in Emory's Rep., t. 5; Gray, Pl. Fendl., p. 93. Main Fork of Red River; June 25—July 8. The pappus is more hyaline and acute than in specimens from other localities in my herbarium. It is also slightly lacerate at the tip, showing something of a transition to *R. arachnoidea.* The leaves, too, are more woolly and broader than in the more common form of the plant.

RUDBECKIA HIRTA, Linn.; Torr. and Gr., Fl. 2, p. 307. Witchita Mountains; fl. June 1. Is *R. bicolor* distinct from this species? Dr. Gray remarks, (Plant. Lindh. 2, p. 227,) that in cultivation, the purple brown of the rays is commonly obsolete or wanting in all the later heads.

ECHINACEA ANGUSTIFOLIA, DC. Prodr. 5, p. 554; Torr. and Gr., Fl. 2, p. 306. Witchita Mountains; June 1.

LEPACHYS COLUMNARIS, Torr. and Gr., Fl. 2, p. 315. *Rudbeckia columnaris*, Pursh, Fl. 2, p. 575. Common on all the tributaries of the Red River; June.

HELIANTHUS PETIOLARIS, Nutt. in Jour. Acad. Philad. 2, p. 115; Sweet Brit. Fl. Gard. (n. ser.) t. 75. With the preceding.

GAILLARDIA PULCHELLA, Foug.; DC. Prodr. 5, p. 652; Torr. and Gr., Fl. 2, p. 366. Common on the upper tributaries of the Red River; May—June.

PALAFOXIA CALLOSA, Torr. and Gr., Fl. 2, p. 369. *Stevia callosa*, Nutt. in Jour. Acad. Philad. 2, p. 121; Bart. Fl. Amer. Sept., t. 46. β. *foliis latioribus.* Tributaries of the Washita; June.

HYMENOPAPPUS CORYMBOSUS, Torr. and Gr., Fl. 2, p. 372. *H. Engelmannianus*, Kunth.

ACTINELLA LINEARIFOLIA, Torr. and Gr., Fl. 2, p. 383. *Hymenoxys linearifolia*, Hook. Witchita Mountains; May 30.

MARSHALLIA CAESPITOSA, Nutt. in DC. Prodr. 5, p. 680; Hook. Bot. Mag. t. 3,704; Torr. and Gr., Fl. 2, p. 391. Headwaters of the Trinity River; May.

APHANOSTEPHUS RAMOSISSIMUS, DC. Prodr. 5, p. 310; Gray, Pl. Wright. 1, p. 93. *A. Riddellii*, Torr. and Gr., Fl. 2, p. 189. *Egletes ramosissima*, Gray, Pl. Fendl., p. 71. Little Witchita and upper tributaries of Red River; May—June. The tube of the disk flowers is indurated in all the specimens.

ENGELMANNIA PINNATIFIDA, Torr. and Gr., in Nutt. Trans. Am. Phil. Soc. (n. ser.) 7, p. 343; and Fl. 2, p. 283. Witchita Mountains; May 30.

MELAMPODIUM CINEREUM, DC. Prodr. 5, p. 518; Gray, Pl. Fendl., p. 78; *M. ramosissimum*, DC. l. c., Torr. and Gr., Fl. 2, p. 271. *M. lencanthum*, Torr. and Gr. l. c. Cache Creek; June 21. A variable species.

CHRYSOPSIS CANESCENS, Torr. and Gr., Fl. 2, p. 256; Gray, Pl. Fendl., p. 77. Main Fork of Red River; July 8.

C. HISPIDA, Hook. Fl. Bor.—Am. 2, p. 22, (under *Diplopappus;*) DC. Prodr. 7, p. 279; Torr. and Gr. l. c.

CENTAUREA AMERICANA, Nutt. in Jour. Acad. Phil. 2, p. 117; Bart. Fl. Amer.—Sept., t. 50; Torr. and Gray, Fl. 2, p. 453. Tributaries of the upper Red River; June—July.

CIRSIUM UNDULATUM, Spreng.; Torr. and Gr., Fl. 2, p. 456. With the preceding.

PYRRHOPAPPUS CAROLINIANUS, DC. Prodr. 7, p. 144; Nutt. in Trans. Amer. Phil Soc. (n. ser.) 7, p. 430. Headwaters of the Trinity and on Cache Creek; May.

LYGODESMIA JUNCEA, Don.; Hook. Fl. Bor.—Am. 2, p. 295, t. 103; Torr. and Gr., Fl. 2, p. 484. Upper tributaries of the Red River; June.

The lower branches are covered at the base with tubers or galls, about the size of cherry-stones, produced by the stings of insects.

L. APHYLLA, DC. Prodr. 7, p. 198; Torr. and Gr., Fl. 2, p. 485. β. *Texana*, Torr. and Gr. l. c. North Fork of Red River; June 16. The numerous radical leaves are 3–4 inches long, runcinately pinnatifid. Achenia angular, distinctly tapering upward.

ASCLEPIADACEÆ.

ASCLEPIAS TUBEROSA, Linn.; Michx. Fl. 1, p. 117; Sweet Brit. Fl. Gard. (ser. 2,) t. 24; Decaisne, in DC. Prodr. 8, p. 567. Torr. Fl. N. York, 2, p. 123. Upper tributaries of Red River; May—June. The leaves vary from ovate and amplexicaul to narrowly linear.

A. SPECIOSA, Torr., in Ann. Lyc. 2, p. 218, and in Fremont's First Rep., p. 95. *A. Douglasii*, Hook. Fl. Bor.–Am. 2, p. 53, t. 142; Decaisne, l. c. Witchita Mountains to the upper tributaries of the Red River; fl. June—July; flowers larger than in any other North American species of Asclepias.

ACERATES PANICULATA, Decaisne, l. c., p. 521; Asclepias viridis, Walt., Fl. Carol. p. 107.? *Anantherix paniculatus*, Nutt., in Trans. Amer. Phil. Soc., (n. ser.,) 5, p. 202. Cache Creek and Middle Fork of Red River; fl. May 16, fr. June.

A. DECUMBENS, Decaisne, l. c. *Anantherix decumbens*, Nutt. l. c. Cache Creek; fl. May 17. The follicles oblong, not muricate.

A. ANGUSTIFOLIA, Decaisne, l. c. *Polyotus angustifolius*, Nutt. l. c. Branch of Cache Creek; fl. May 17.

A. VIRIDIFLORA, Ell. sk. 1, p. 317; Torr. Fl. N. York, 2, p. 124; Decaisne, l. c. *Asclepias viridiflora*, Pursh, Fl. 1, p. 181; Hook. Fl. Bor.–Am. 2, p. 53, t. 143. North Fork of Red River; fl. June 4. The specimens collected by Captain Marcy belong to the broad-leaved forms of the plant.

ENSLENIA ALBIDA, Nutt. Gen. 1, p. 164, and in Trans. Amer. Phil. Soc., (n. ser.) 5, p. 203; Decaisne, in DC. Prodr. 8, p. 518. Main Fork of Red River; not in flower.

APOCYNACEÆ.

APOCYNUM CANNABINUM, Linn.; Hook. Fl. Bor.—Amer. 2, p. 51, t. 139; Decaisne, in DC. Prodr. 8, p. 439; Torr. Fl. New York, 2, p. —. Common on the upper tributaries of Red River; May—June.

AMSONIA SALICIFOLIA, Pursh, Fl. 1, p. 184; Decaisne, in DC. Prodr. 8, p. 385. Witchita Mountains; fr. July 16. This is perhaps only a variety of *A. angustifolia*, Michx., and both may not be specifically distinct from *A. tabernæmontana*.

GENTIANACEÆ.

SABBATIA CAMPESTRIS, Nutt., in Trans. Amer. Phil. Soc., (n. ser.,) 5, p. 167; Griseb., in DC. Prodr. 9, p. 50; Engelm. and Gr., Pl. Lindh. 1, p. 15. On the Washita; fl. and fr. July 27.

ERYTHRÆA BEYRICHII, Torr. and Gr., Fl. 2, ined. *E. trichantha β. angustifolia*, Griseb. l. c. With the preceding; fl. and fr. July 26.

EUSTOMA RUSSELIANUM, Don.; Griseb. in DC. Prodr. 8, p. 51. *Lisianthus glaucifolius*, Nutt. l. c. *L. Russelianus*, Hook. Bot. Mag., t. 3626. Washita River to the upper tributaries of the Red River; July.

CONVOLVULACEÆ.

EVOLVULUS PILOSUS, Nutt. Gen. 1, p. 174, (as a synonym); Trans. Amer. Phil. Soc., (n. ser.,) 5, p. 195. *E. argenteus*, Pursh, Fl. 1, p. 187; Choisy, in DC. Prodr. 9, p. 443; not of R. Br. Middle Fork of Red River; fl. May 22. Choisy doubtingly refers Brown's plant to *E. hirsutus*, Lam., and therefore has adopted Pursh's name.

CONVOLVULUS LOBATUS, Engelm., and Gray, Pl. Lindh. 1, p. 44 (in a note.) *C. hastatus*, Nutt., in Trans. Amer. Phil. Soc., (n. ser.,) 5, p. 194; not of Thunb. *C. Nuttallii*, Torr. in Emory's Rep., p. 149. Middle Fork of Red River; May 22—June 6. This species has much the appearance of *C. althæoides*, Boss.

C. (IPOMŒA) LEPTOPHYLLUS, Torr., in Frem. First Report, p. 94, and in Emory's Report, p. 148, t. 11. With the preceding.

C. (IPOMŒA) SHUMARDIANUS, (sp. nov.;) caule gracili subpubescente; foliis ovato-lanceolatis sursum angustatis basi acutis; pedunculis petiolas longioribus 2-4-floris; sepalis ovatis obtusis. Witchita Mountains; fl. July 17; flowers as large as in *C. panduratus*, which the plant much resembles, but differs in the form of the leaves, and in the broader and more obtuse sepals. Named in honor of Dr. G. C. Shumard, the botanical collector of the expedition.

SOLANACEÆ.

SOLANUM FLAVIDUM, Torr. Ann. Lyc. New York, 2, p. 227; Dunal in DC. Prodr. 13, p. 375. Cache Creek; May. This species is not suffrutescent, as is stated in the original description, but probably annual. Mr. Wright found it on the Rio Grande. The prickles are sometimes almost wanting.

S. CAROLINENSE, Linn.; Torr., Fl. N. York 2, p. 105; Dunal, l. c., p. 305. Witchita Mountains and upper tributaries of the Red River; May–June.

PHYSALIS PUMILA, Nutt., in Trans. Amer. Phil. Soc., (n. ser.,) 5, p. 193. With the preceding; May–June. This species has been overlooked by Dunal in DC. Prodr.

SCROPHULARIACEÆ.

CASTILLEJA PURPUREA, G. Don.; Benth., in DC. Prodr. 10, p. 531. *Euchroma purpurea*, Nutt., l. c., p. 180. Sources of the Trinity River; May.

PENTSTEMON GRANDIFLORUS, Nutt., in Fras. Cat. 1813, and Gen. 2, p. 53; Benth., l. c., p. 322. *P. Bradburii*, Pursh, Fl. 2, p. 738. North Fork of Red River; fl. June 3. The pedicels vary from three lines to nearly an inch in length.

P. AMBIGUUS, Torr., in Ann. Lyc. N. York, 2, p. 228; Benth., l. c., p. 321. Witchita Mountains; June. This rare and well characterized species has lately been found by Mr. Wright on the Upper Rio Grande.

P. COBÆA, Nutt., l. c.; Hook. Bot. Mag., t. 3465; Benth., l. c., p. 326. Upper tributaries of the Red River; May–June.

P. PUBESCENS, Soland.; Torr., Fl. N. York, 2, p. 35; Benth., l. c. Headwaters of the Trinity. Smoothish, with narrower and more entire leaves than usual.

GERARDIA GRANDIFLORA, Benth., Comp. Bot. Mag., 1, p. 206. *Dasystoma Drummondi,* Benth., in DC. Prodr. 10, p. 521. On the Washita; fl. July 27.

LABIATÆ.

MONARDA ARISTATA, Nutt., in Trans. Amer. Phil. Soc., (n. ser.,) 5, p. 186; Benth., in DC. Prodr. 12, p. 363. Main Fork of Red River; May 24–25. Nuttall says that this species is sometimes perennial; but all our specimens seem to be annual. A variety was found on Cache Creek, in which the teeth of the calyx are aristate from a broad base, and strongly hispid-ciliate. The corolla is not spotted, as in the ordinary form.

M. PUNCTATA, Linn.; Benth., l. c.; Torr., Fl. N. York, 2, p. 59. *M. lutea,* Michx., Fl. 1, p. 16. North and Middle Forks of Red River; May–June. A dwarfish and annual form, in which the corolla is scarcely spotted, was found in the same region.

TEUCRIUM CUBENSE, Linn.; Benth., in DC. Prodr. 12, p. 578. *T. laciniatum,* Torr., in Ann. Lyc. New York, 2, p. 231. Cache Creek and Middle Fork of Red River; May. This species was incorrectly described by me as "fruticulose" in the work quoted.

SCUTELLARIA RESINOSA, Torr., in Ann. Lyc. N. York, 2, p. 232; Benth., in DC. Prodr. 12, p. 427. Cache Creek and Sweetwater Creek; May 18–June 9.

S. PARVULA, Michx., Fl. 1, p. 12; Benth., l. c.; Torr., Fl. N. York, 2, p. 71. *S. ambigua,* Nutt., Gen. 2, p. 37.

VERBENACEÆ.

LIPPIA CUNEIFOLIA, Torr., in Ann. Lyc. N. York, 2, p. 234, (under *Zapania.*) Witchita Mountains, and on the Washita; June 1–27. Schauer has overlooked this species, in his revision of *Verbenaceæ* for DC. Prodr.

VERBENA BIPINNATIFIDA, Engelm. and Gray, Pl. Lindh. 1, p. 49; Schauer, in DC. Prodr. 11, p. 553. *Glandularia bipinnatifida,* Nutt., in Jour. Acad. Phil. 2, p. 123, and in Amer. Phil. Trans. (n. ser.) 5, p. 184. Sources of the Trinity and upper tributaries' of Red River; May–June.

BORAGINACEÆ.

EUPLOCA CONVOLVULACEA, Nutt., in Amer. Phil. Trans., (n. ser.) 5, p. 190; DC. Prodr. 9, p. 559. Middle Fork of Red River; fl. June 23. I am now convinced that my *E. grandiflora* (Emory's Report, p. 147) is an unusually large-flowered state of the present species. The plant is abundant on the Upper Rio Grande.

ERITRICHIUM JAMESII. *Myosotis suffruticosa*, Torr., in Ann. Lyc. N. York, 2, p. 225; DC. Prodr. 10, p. 114. North Fork of Red River; fl. and fr. June 14. This plant had not been found, till Captain Marcy collected it, since it was discovered by Dr. James, in Long's Expedition. It is a genuine *Eritrichium*, but can hardly be referred to any one of De Candolle's sections of that genus. My description (l. c.) was drawn from old and imperfect specimens, the stems of which were indurated at the base so as to appear suffrutescent. As more complete specimens show the plant to be herbaceous, the former specific name is not appropriate. The allied Fendlerian species No. 636 (*E. multicaule* Torr. Mss.) is very hispid and canescent, with spreading hairs, and throws up several stems from a thick root or caudex. Leaves linear-spatulate and obtuse. Flowers on conspicuous pedicels. Fructiferous calyx broadly ovate, nearly erect; the segments ovate-lanceolate and closed over the fruit. Nutlets truncate at the summit, very smooth and shining.

POLEMONIACEÆ.

PHLOX PILOSA, Linn.; Benth., in DC. Prodr. 9, p. 305. Sources of the Trinity; May.

PRIMULACEÆ.

DODECATHEON MEADIA, Linn.; Pursh, Fl. 1, p. 136; DC. Prodr. 8, p. 56. Sources of the Trinity; fl. May.

SANTALACEÆ.

COMANDRA UMBELLATA, Nutt. Gen 1, p. 157; Hook. Fl. Bor.–Am. 2, p. 139, t. 79, f. A; Torr. Fl. N. York, 2, p. 160. *Thesium umbellatum*, Linn. Tributaries of the Red River; May. There are few plants that have a wider range in latitude and longitude than this.

EUPHORBIACEÆ.

EUPHORBIA COROLLATA, Linn.; Pursh, Fl. 2, p. 607; Torr. Fl. N. York, 2, p. 175, t. 99. On the Washita; July.

E. MARGINATA, Pursh, Fl. 2, p. 607; Torr. in Ann. Lyc. N. York, 2, p. 224. Main Fork of Red River; July 8. Upper part of the stem hairy.

E. HELIOSCOPIA, Linn.; Torr. Fl. N. York, 2, p. 174, (excl. syn. Pursh;) Gray, Bot. N. States, p. 405. Headwaters of the Trinity; fl. May.

STILLINGIA LANCEOLATA, Nutt. in Trans. Amer. Phil. Soc., (n. ser.) 5, p. 176. *S. sylvatica β. salicifolia*, Torr. in Ann. Lyc. N. York, 2, p. 245. Middle Fork of Red River; fl. June 4.

HENDECANDRA TEXENSIS, Klotsch in Erich. Arch. (1841) 1, p. 252; Engel. and Gray, Pl. Lindh. 1, p. 53. *Croton muricatum*, Nutt. in Trans. Amer. Phil. Soc. (n. ser.) 5, p. 153. *H. multiflora*, Torr. in Frem. First Rep., p. 96. Middle Fork of Red River; fl. and fr. June 22.

GYNAMBLOSIS MONANTHOGYNA. *Engelmannia Nuttalliana*, Klotsch, l. c. *Croton monanthogynum*, Michx. Fl. 2, p. 215. *C. ellipticum*, Nutt. Gen. 2, p. 225, (excl. syn.;) Torr. in Ann. Lyc. N. York, 2, p. 245. Main Fork of Red River; June 24. The *Engelmannia* of Klotsch, which is based on *Croton ellipticum* of Nuttall, must give place to the earlier genus of the same name of Torr. and Gray. I propose for it a manuscript name given to the plant many years ago, when revising the *Euphorbiaceæ* of the United States. Klotsch is wrong in referring *Croton monanthogynum* to *Hendecandra maritima*. In the young specimens of Captain Marcy all the staminate flowers are 8–10 androus: and the later flowers are not unfrequently hexandrous. The petals and sepals vary from three to five.

TRAGIA RAMOSA, Torr. in Ann. Lyc. N. York, 2, p. 245. *T. angustifolia*, Nutt., in Trans. Amer. Phil. Soc. (n. ser.) 5, p. 172. *T. brevispica*, Engel. and Gray, Pl. Lindh. 1, p. 54. North Fork of the Red River; June.

CNIDOSCOLUS STIMULOSUS, Engel. and Gray, Pl. Lindh. 1, p. 26. *Jatropha stimulosa*, Michx. Fl. 2, p. 216; Ell. Sk. 2, p. 649. Cache Creek; May 17.

PLANTAGENACEÆ.

PLANTAGO VIRGINICA, Linn.; Torr. Fl. New York, 2, p. 16. Headwaters of the Trinity; fl. May.

P. GNAPHALOIDES, Nutt. Gen. 1, p. 100; Hook. Fl. Bor.—Am. 2, p. 124; Decaisne in DC. Prodr. 13, (Saet. 1,) p. 713. Mouth of the Big Medicine River.

POLYGONACEÆ.

ERIOGONUM LONGIFOLIUM, Nutt. in Trans. Amer. Phil. Soc. (n. ser.) 5, p. 164; Benth. Eriog. in Linn. Trans. 17, p. 406. Witchita Mountains; June.

CHENOPODIACEÆ.

CHENOPODIUM SUBSPICATUM, Nutt. Gen. 1, p. 199? Middle Fork of Red River. The specimens are without either flowers or fruit. Annual, diffuse, and much branched; clothed with whitish furfuraceous scales. Leaves conspicuously petiolate, broadly rhombic-ovate, with one or two coarse teeth on each side.

OBIONE CANESCENS, Moq. Chenop., p. 74; and in DC. Prodr. 13, (pars 2,) p. 113; Torr., in Stansbury's Report, p. 395. *O. occidentalis*, Moq. l. c. *Calligonium canescens*, Pursh, Fl. 2, p. 370. *Atriplex canescens*, Nutt. Gen. 1, p. 197. Common on the upper tributary of the Red River.

NYCTAGINACEÆ.

OXYBAPHUS ANGUSTIFOLIUS, Torr., in Ann. Lyc. N. York, 2, p. 237; Choisy, in DC. Prodr. 13, (pars 2,) p. 433. *Calymenia angustifolia*, Nutt., in Fras. Cat. 1813, and Gen. 1, p. 26. Upper tributaries of Red River; June.

O. NYCTAGINEUS, Torr. l. c.; Choisy, l. c. *Allionia nyctaginea*, Michx., Fl. 1, p. 100. *Calymenia corymbosa*, Nutt., in Trans. Amer. Phil. Soc., (n. ser.) 5, p. 178; not *Mirabilis corymbosa*, Cav., in which the involucrum is one-flowered. With the preceding; May 28.

O. HIRSUTUS, Sweet; Hook. Fl. Bor.-Amer. 2, p. 124; Choisy, l. c. *Allionia hirsuta*, Pursh, Fl. 2, p. 728. With the preceding. ♃. Stem erect, 2-3 feet high, sparingly branched; viscously pubescent; leaves 2-3 inches long and 1-1½ inch wide, on very short petioles, nearly entire. Flowers in a long, loose terminal and naked panicle; involucre 3-flowered, rotate-companulate. Fruit fusiform, oblong, 5-angled. As in most of the *Nyctaginaceæ*, this plant abounds in cells filled with raphides. These are so abundant in the liber of the root, that they forma layer of a silvery white color.

ABRONIA MELLIFERA, Dougl., in Hook. Bot. Mag., t. 2879; Choisy, l. c. Cache Creek; fl. and fr. May 18. The specimens in the collection agree exactly with Douglas's plant collected in California, and named by Sir William Hooker.

CUPULIFERÆ.

QUERCUS UNDULATA, Torr., in Ann. Lyc., 2, p. 248, t. 4. Abundant on the upper tributaries of the Red River. Stems 1-2 feet long, from a thick woody base, sparingly branched above. Leaves oblong, two inches or more in length, undulate, and furnished with 1-3 rather obtuse and scarcely mucronate teeth on each side, densely and softly pubescent underneath, nearly smooth above, thick and somewhat coriaceous.

CONIFERÆ.

JUNIPERUS VIRGINIANA, Linn.; Michx. f. Sylv. 2, p. 353, t. 155; Torr., Fl. N. York, 2, p. 235. *J. sabina*, Hook., Fl. Bor.-Am. 2, p. 166. Middle Fork of Red River.

HYPOXIDACEÆ.

HYPOXIS ERECTA, Linn.; Bart., Fl. N. Amer, 1, t. 35, f. 1; Torr., Fl. N. York, 2, p. 289. Headwaters of the Trinity River; May.

COMMELYNACEÆ.

COMMELYNA ANGUSTIFOLIA, Linn.; Kunth, Enum. 4, p. 53; Torr., Fl. N. York, 2, p. 332. North Fork of Red River; May-June.

TRADESCANTIA VIRGINICA, Linn.; Bot. Mag., t. 105; Bart. l. c., t. 41; Kunth, Enum. 4, p. 81; Torr., Fl. N. York, 2, p. 333. Abundant on the upper tributaries of Red River; May–June; extremely variable in pubescence, and in the breadth of the leaves.

IRIDACEÆ.

SISYRINCHIUM BERMUDIANA, Linn.; Torr., Fl. N. York, 2, p. 290. Headwaters of the Trinity; May.

NEMASTYLIS ACUTA; with the preceding.

LILIACEÆ.

SCILLA ESCULENTA, Ker. Bot. Mag., t. 1574. *Phalangium esculentum*, Nutt., in Fras. Cat. 1813, Gen. 1, p. 219. *P. Quamash*, Pursh, Fl. 1, p. 226. Headwaters of the Trinity; May.

ALLIUM CANADENSE, Kalm; Pursh, Fl. 1, p. 223; Kunth, Enum. 4, p. 450; Torr., Fl. N. York, 2, p. 308. On Cache Creek; fl. May 14.

A. OCHROLEUCUM, Nutt. Trans. Amer. Phil. Soc. (n. ser.) 5, p. 156; not of Waldst. and Kit. Headwaters of the Trinity; May.

A. RETICULATUM, Fras.? Kunth, Enum. 4, p. 435. *A. angulosum, β. lenchorhizum*, Nutt. l. c.? Common on the tributaries of Red River. Bulb usually covered with dark reticulated coats, but sometimes naked.

MELANTHACEÆ.

AMIANTHIUM NUTTALII, Gray, in Ann. Lyc. N. York, 4, p. 123. *Helonias angustifolia*, Nutt., in Trans. Amer. Phil. Soc., (n. ser.) 5, p. 154. *Amiantanthus*, Kunth, Enum. 4, p. 181. Headwaters of the Trinity; May.

CYPERACEÆ.

CYPERUS SCHWEINITZII, Torr. Cyp., p. 276; Fl. N. York, 2, p. 343. *C. alterniflorus*, Schwein., in Long's 2d Exped., 2, p. 381, (not of R. Br.) Middle and North Forks of Red River; May–June.

C. STRIGOSUS, Linn.; Torr. Cyp., p. 261; Fl. N. York, 2, p. 340, t. 136. Witchita Mountains; July.

CYPERUS ACUMINATUS, Torr. and Hook., in Torr. Cyp. Suppl. Witchita Mountains; July 15.

FIMBRISTYLIS SPADICEA, Vahl, Enum. 2, p. 294; Torr. Cyp., p. 346; Kunth, Enum. 2, p. 237; Torr. Fl. N. York, 2, p. 360. Headwaters of the Trinity; May.

CAREX MUHLENBERGII, Schk. Car. 2, p. 12, f. 178; Schwein. and Torr. Car., p. 304; Torr. Fl. N. York, 2, p. 374. Headwaters of the Trinity, and on Cache Creek; May–June.

C. FESTUCACEA, Schk. Car. f. 173; Carey, in Gray's Bot. N. States, p. 545. *C. straminea*, var. *festucacea*, Torr. l. c. With the preceding.

GRAMINEÆ.

PHALARIS ANGUSTA, Nees; Trin. Ic. Gram. t. 78; Kunth, Gram. 2, p. 32. *P. occidentalis*, Nutt., in Trans. Amer. Phil. Soc., (n, ser.) 5, p. 144. On Cache Creek; May 16. This plant is certainly *P. angusta* of Trinius, of which I have specimens named by that distinguished botanist. It appears, however, scarcely to differ from *P. microstaclya*, DC.

PASPALUM LÆVE, Michx. Fl. 1, p. 44; Trin. Panic. Gen., p. 160; Torr., Fl. N. York, 2, p. 421. Main Fork of Red River; June.

PANICUM PAUCIFLORUM, Ell. Sk. 1, p. 120; Gray, Bot. N. States, p. 613. Headwaters of the Trinity; May.

P. RETICULATUM, (n. sp.;) culmo geniculato erecto subsimplici; foliis vaginisque laxe pilosis; panicula oblonga contracta, ramulis racemosis paucifloris; spiculis obovatis acutiusculis glabris breviter pedicellatis muticis; glumis valde inæqualibus; flore neutro bivalvi; palea inferiore (ut in gluma superiora) 7-costulata reticulata, flore hermaphrodito transverse ruguloso. On the Main Fork of Red River; July. Nos. 2090 and 2091, Wright's Coll. N. Mex. 1851–52, are glabrous and more robust forms of this species.

P. OBTUSUM, (H. B. K.?) spicis 5-7 racemosim dispositis erectis; spiculis geminis subimbricatis unilateralibus muticis obovatis obtusis glabris; glumis æqualibus multinervosis; flore inferiore triandro bipaleaceo; flore hermaphrodito subtilissime longitudinaliter striato subnitido.—H. B. and Kunth, Nov. Gen. 1, p. 98? Tributaries of the Washita. Plant glabrous and glaucous, about 18 inches long. Rachis narrowly linear, very flexuous; nerves of the glumes green. Near *P. obtusum*,

H. B. K., but differs in the nearly equal glumes, &c. No. 2092 Wright's Coll. N. Mex. 1851–52, is exactly our plant.

ARISTIDA FASCICULATA, Torr., in Ann. Lyc. N. York, 2, p. 154; Kunth, Enum. 2, p. 196. *A. purpurea*, Nutt. in Trans. Amer. Phil. Soc., (n. ser.) 5, p. 145. Middle Fork of Red River; May—June.

AGROSTIS (SPOROBOLUS) AIROIDES, Torr., in Ann. Lyc. N. York, 2, p. 151. With the preceding. The axils of the panicle are nearly glabrous in Captain Marcy's specimens.

CALAMAGROSTIS GIGANTEA, Nutt. l. c., p. 143. Middle Fork of Red River; June 23.

CHLORIS VERTICILLATA, Nutt. l. c. With the preceding; June 25. An elegant grass, near *C. alba*, Presl. and Torr. in Emory's Rep., p. 153.

BOUTELOUA RACEMOSA, Lag. Var. Cienc. (1805) p. 141; Torr. in Emory's Rep., p. 154; not of Torr. Fl. N. York. *Dinebra curtipendula*, DC.? Kunth, Syn. Pl. Eq. 1, p. 281; excl. syn. Michx. *Eutriana curtipendula*, Trin. Fund. p. 161 (in part); Kunth, Enum. 1, p. 280, and Suppl. p. 233; excl. syn. Michx. and Willd. Main Fork of Red River; July. The detailed description of this species by Kunth, l. c., (drawn from a Mexican specimen collected by Humboldt) shows that the *Chloris curtipendula* of Michaux (*Bouteloua curtipendula*, Torr.) is a distinct species, as indicated in Emory's Report, l. c.

CHONDROSIUM OLIGOSTACHYUM. *Atheropogon oligostachyum*, Nutt. Gen. 1, p. 78; Torr. in Sill. Jour. 4, p. 58. *Eutriana? oligostachya*, Kunth, Gram. 1, p. 96, ex. Enum. 2, p. 282. Main Fork of Red River; July 2.

C. PAPILLOSUM. *Atheropogon papillosum*, Engelm. in Sill. Jour. 46, p. 104. With the preceding, of which it is perhaps only a variety. The species of *Chondrosium* and *Bouteloua* are known by the name of *Grama Grasses* in New Mexico and Texas.

PLEURAPHIS JAMESII, Torr. in Ann. Lyc. N. York, 1, p. 148, t. 10; Kunth, Enum. 1, p. 285. Main Fork of Red River; July. Kunth (l. c.) asks whether this is not *Hymenothecium quinquesetum* of Lagasca; but the brief description of that author (in Gen. et Sp. Pl. Nov. 1816) does not agree with our plant.

SESLERIA DACTYLOIDES, Nutt. Gen. 1, p. 65; Kunth, Enum. 1, p. 323; Torr. in Emory's Report, p. 323, t. 10. Upper tributaries of the Red River; July. This is the well known Buffalo-grass of the western prairies. It is remarkable that neither the grain nor the fertile flowers of this grass are known.

POA (ERAGROSTIS) OXYLEPIS. *P. interrupta*, Nutt. in Trans. Amer. Phil. Soc., (n. ser.) 5, p. 146; not of Lam. Witchita Mountains; July. A very neat grass. The specimens of Captain Marcy are only about 18 inches high.

P. ERAGROSTIS, Linn.; Kunth, Enum. 1, p. 333; Torr., Fl. N. York, 2, p. 458. North Fork of Red River; July.

P. ARACHNIFERA: panicula oblonga contracta, ramulis semiverticillatis; spicis subquinquefloris, lato-ovatis, floribus laxis basi et racheos longe lanoso-arachnoideis; glumis inequalibus anguste-lanceolatis, in carina scabris; palea inferiore lineari-lanceolata acutissima obscure 3-5-nervata, carina inferne ciliata.

β? spiculis 9-10 floris, rachi sparsa lanosa. Headwaters of the Trinity; May.

MELICA GLABRA, Michx. Fl. 1, p. 62. Witchita Mountains; May 30.

KOELERIA CRISTATA, Pers. Syn. 1, p. 97; Kunth, Enum. 1, p. 381. *K. nitida*, Nutt. Gen. 1, p. 74. *K. tuberosa*, Nutt. in Amer. Phil. Trans. (n. ser.) 5, p. 148. Headwaters of the Trinity.

FESTUCA NUTANS, Willd. Enum. 1, p. 116; Kunth, Enum. 1, p. 407; Torr. Fl. N. York, 2, p. 471, t. 158. Witchita Mountains; June.

F. TENELLA, Willd. 1 c.; Kunth, Enum. 1, p. 397; Torr., Fl. N. York, 2, p. 470, t. 154. Headwaters of the Trinity: taller than the plant of the Atlantic States.

UNIOLA LATIFOLIA, Michx., Fl. 1, p. 71; Ell. Sk. 1, p. 167; Kunth, Enum. 1, p. 425. Witchita Mountains; July. A tall, showy grass, with very large much compressed spikelets.

U. STRICTA, Torr., in Ann. Lyc. N. York, 1, p. 155. *U. multiflora*, Nutt., in Trans. Amer. Phil. Soc. (n. ser.) 5, p. 148. Washita River to the upper tributaries of the Red River; June–July. No. 2033 Wright's Coll. N. Mex. 1851–52 is the same.

TRITICUM REPENS, Linn.; Kunth, Enum. 1, p. 440; Torr., Fl. N. York, 2, p. 474. Common on the tributaries of Red River; May–June. All the specimens are awnless.

ELYMUS CANADENSIS, Linn.; Kunth, Enum. 1, p. 451; Torr., Fl. N. York, 2, p. 476. *E. glaucifolius*, Willd. Cache Creek, &c.; June.

HORDEUM JUBATUM, Linn.; Torr., Fl. Mid. and N. States, 1, p. 158; Kunth, Enum. 1, p. 457. Tributaries of Red River.

H. PUSILLUM, Nutt. Gen. 1, p. 87, and Trans. Amer. Phil. Soc. (n. ser.) 5, p. 151; Kunth, Enum. 1, p. 457.

TRIPSACUM DACTYLOIDES, Linn; Michx. Fl. 1, p. 61; Nutt. l. c.; Kunth, Enum. 1, p. 469. North Fork of Red River; June.

ANDROPOGON JAMESII. *A. glaucum*, Torr., in Ann. Lyc. N. York, 1, p. 153; not of Muhl. With the preceding.

EQUISETACEÆ.

EQUISETUM HYEMALE, Linn.; Pursh, Fl. 2, p. 652; Torr., Fl. New York, 2, p. 482. Main Fork of Red River.

EXPLANATION OF PLATES.

Plate I. ANEMONE CAROLINIANA.
    Fig. 1, a stamen, magnified; fig. 2, a head of pistils; fig. 3, a head of ripe achenia, both magnified; fig. 4, a single achenium, more enlarged.

Plate II. DITHYRÆA WISLIZENI.
    Fig. 1, a flower, magnified; fig. 2, the pistil, more enlarged; fig. 3, a ripe pod, with one cell opened, to show the seed—also magnified; fig. 4, the embryo, more magnified.

Plate III. GERANIUM FREMONTII.*

Plate IV. HOFFMANSEGGIA JAMESII.
    Fig. 1, a flower; fig. 2, a pod; fig. 3, seed—all moderately magnified.

Plate V. SANGUISORBA ANNUA.
    Fig. 1, a flower; fig. 2, the fruit—both magnified.

Plate VI. ERYNGIUM DIFFUSUM.
    Fig. 1, a separate leaf; fig. 2, a flower; fig. 3, a petal; fig. 4, the ovary, with the styles and three of the sepals; fig. 5, front view of a stamen and sepal; fig. 6, side view of the same—all but fig. 1 more or less magnified.

Plate VII. EURYTÆNIA TEXANA.
    Fig. 1, a mericarp, magnified; fig. 2, transverse section of the same, more magnified.

Plate VIII. LIATRIS ACIDOTA.
    Fig. 1, head of flowers, moderately magnified; fig. 2, a single flower, more enlarged; fig. 3, a single bristle of the pappus, still more enlarged.

Plate IX. APHANOSTEPHUS RAMOSISSIMUS.
    Fig. 1, a ray-flower; fig. 2, a disk-flower; fig. 3, style of the same; fig. 4, achenium, with its coroniform pappus—all magnified.

Plate X. XANTHISMA TEXANA.
    Fig. 1-3, scales of the involucre; fig. 4, a disk-flower; fig. 5, achenium and pappus of the same; fig. 6, ray-flower; fig. 7, style of the disk-flower—all magnified.

---

* This species was not found by Captain Marcy, but it grows in the region that he explored. The plate was prepared for another government report, which was never published.

Plate XI. ENGELMANNIA PINNATIFIDA.

Fig. 1, a ray-flower, with an inner involucral scale; fig. 2, style of the same; fig. 3, a disk-flower; fig. 4, style of the same; fig. 5, an achenium—all magnified.

Plate XII. ARTIMESIA FILIFOLIA.

Fig. 1, portion of a flowering branch, moderately enlarged; fig. 2, a single head, more magnified; fig. 3, the same, longitudinally cut and equally magnified; fig. 4, a disk-flower, and fig. 5, a ray-flower, both more magnified.

Plate XIII. ERYTHRÆA BEYRICHII.

Fig. 1, a flower, magnified; fig. 2, a capsule.

Plate XIV. HELIOTROPIUM TENELLUM.

Fig. 1, the calyx; fig. 2, corolla, showing its æstivation; fig. 3, the same, expanded; fig. 4, the same, laid open; fig. 5, fruit; fig. 6, longitudinal section of the seed—all magnified.

Plate XV. EUPLOCA CONVOLVULACEA.

Fig. 1, a flower, moderately magnified; fig. 2, the same, aid open and equally magnified; fig. 3, the stamens, more magnified; fig. 4, a single stamen, still more magnified; fig. 5, the pistil, equally magnified; fig. 6, fruit, with the persistent style; fig. 7, transverse section of the same, equally enlarged; fig. 7, longitudinal section of a seed, more magnified.

Plate XVI. PENTSTEMON AMBIGUUS.

Fig. 1, a flower, moderately magnified; fig. 2, the stamens and a portion of the corolla, more enlarged; fig. 3, the pistil, equally magnified; fig. 4, capsule, twice the natural size, and dehiscent.

Plate XVII. LIPPIA CUNEIFOLIA.

Fig. 1, a bract; fig. 2, a flower; fig. 3, the calyx; fig. 4, the corolla, cut longitudinally, showing the stamens and pistil—all moderately magnified; fig. 5, the pistil, longitudinally cut, more enlarged.

Plate XVIII. ABRONIA CYCLOPTERA.

Fig. 1, involucre, somewhat magnified; fig. 2, fruit of the natural size; fig. 3, transverse section of the fruit, magnified; fig. 4, an achenium, magnified; fig. 5, transverse section of the same, also magnified; fig. 6, the embryo.

Plate XIX. POA INTERRUPTA.

Fig. 1, a spikelet; fig. 2, a single flower; fig 3, a caryopsis—all magnified.

Plate XX. UNIOLA STRICTA.

Fig. 1, a spikelet, magnified.

APPENDIX H.

ETHNOLOGY.

VOCABULARIES OF THE COMANCHES AND WITCHITAS, BY CAPT R. B. MARCY; WITH SOME GENERAL REMARKS, BY PROF. W. W. TURNER.

APPENDIX H.

ETHNOLOGY.

VOCABULARIES OF WORDS IN THE LANGUAGES OF THE COMANCHES AND WITCHITAS: BY CAPT. R. B. MARCY.

ENGLISH.	COMANCHE.	WITCHITA.
Man,	To-e-*bitch*-e,	Two-bear-e-*kets*-ah.
Woman,	Wy-e-pe,	*Kah*-haak.
White man,	To-e-*titch*-e,	E-*ka*-rish.
Mexican,	Tack-o-*ti*-bo,	Es-*ta*-he.
Negro,	Toosh-ah-*ty*-bo,	Es-tah-he-*es*-co-rash.
Indian,		*Eh*-hos.
Delaware,		Nar-*wah*-ro.
Kickapoo,		Shake-*kah*-quah.
Cherokee,		*Shan*-nack.
Osage,	Wash-sashe,	*Wash*-sashe.
Comanche,		*No*-taw.
Chief,	Taak-*quin*-no,	A-*ra*-oh.
Friend,	Hartch,	Harteh.
Enemy,	To-ho-*ba*-kah,	Now-*ta*-wah.
One,		*Cha*-osth.
Two,		Witch.
Three,		*Taw*-way.
Four,		*Taalk*-witch.
Five,		Es-*quaw*-etch.
Six,		*Ke*-hash.
Seven,		Ke-*off*-itch.
Eight,		Ke-o-*taw*-wah.
Nine,		Sa-o-*kin*-te.
Ten,		Es-kir-ri-*ah*-wash.
Horse,	Pooke,	Ca-*wah*-ra.
Mule,	*Moo*-rur,	*Moo*-rur.
Bear,	*Whee*-lah,	Wee-rah.
Dog,	*Charl*-lee,	*Keetch*-ah.
Prairie-dog,	*Kee*-chee,	*Keeche*-n'ah.
Sun,	*Tah*-arpe,	*Kee*-shaw.
Moon,	Mushe,	Moir (like French).
Stars,	*Ta*-arche,	Eck-qua-*de*-co.
Water,	Pah,	Keetche.
Fire,	Koo-*o*-nah,	*Es*-tore.
Road,		To-yah-*atch*-co.

ENGLISH.	COMANCHE.	WITCHITA.
Smoke,	*Cook*-toe,	Etch-qua-*ask*-co.
River,	Ho-no,	Hat.
Mountain,	*To*-yah-vees-tah,	Ne-yaw-*caw*-tee.
Corn,	Hah-ne-*be*-tah,	Tais.
Grass,	Me-*cheese*-ka,	Ec-*yock*-cod.
Tree,	*Oho*-pee,	Cawk.
Blanket,	*Wah*-nopp,	Ah-*water*-cotsh.
Mirror,	*Nah*-bo-ne,	Atch-e-*o*-wash.
Paint,	*Pees*-ah-pee,	Tah-rah-*o*-way.
Tobacco,	*Pah*-mo,	*Way*-co.
Powder,	*Nah*-co-chee,	*Eteh*-cod.
Gun,	Pe-*i*-it.	Kah-*to*-kash.
Bow,	Ho-a-*ā*-te,	Kee-*sti*-its.
Arrow,	*Pa*-ark,	*Nay*-quats.
Yes,	Hah,	Wash.
No,	Kay,	Ke-*ah*-re.
To hear,	*Nah*-gut,	To-*otch*-kash.
To sleep,	*Ithe*-pe,	A-shotch-a-*show*-bick.
To come,	Keem-mah,	To-*ta*-os.
To go,	Me-ah-lo,	*Totch*-esch.
Fight,	Naw-bah-*da*-kah,	Ta-*a*-chots.
Understand,	*Hock*-kun-nee,	Wah-tah-chow-*otch*-kash.
Talk,	*Ta*-quaw,	Wash-talk-*ke*-shaw.
Look here,	Cab-*boon*,	Esh-*sha*-esh.
I see,		Un-*sha*-esh.
Tell them,	Marry-e-ah-whit-to,	*E*-shock.
He says,		*Talk*-kash.
How much?		*Atch*-kinch.
How far?		Ah-she-ka-atch-e-*a*-wah.
Good,	Chaat,	*Atch*-tah.
Bad,	*Tahe*-chit,	Naw-*out*-ta.
Great,	*Pe*-opp,	*Totch*-tah.
Small,	*Ter*-titche,	Kee-*etch*-tah.
Black,	*Too*-hop,	*Co*-rash.
Dead,	*Ta*-yeh,	Wah-*ta*-tash.
God,	Tar-*a*-pe.	
My father,	Ner-*ack*-pee.	
My mother,	Ner-*be*-ar.	
My brother,	Ner-*ta*-ma.	
My sister,	Ner-*pa*-cher.	
My son,	Ner-*too*-ar.	
My daughter,	Ner-*pa*-tar.	
My husband,	Ner-co-*mack*-pe.	

APPENDIX H.—ETHNOLOGY. 309

ENGLISH.	COMANCHE.	WITCHITA.
My wife,	Ner-quer.	
Child,	To-*a*-chee.	
Boy,	To-a-*nick*-pe.	
Girl,	Wy-ah-*pee*-chee.	
Face,	*K*oo-veh.	
Body,	*Wahk* cher.	
Head,	*Pa*-aft.	
Heart,	*Pe*-hee.	
Breast,	*To*-koo.	
Hair,	*Par*-pe.	
Hand,	*Moo*-wah.	
Leg,	Ah-*too* koo.	
Foot,	*Nah*-hap.	
Neck,	*Too*-yock.	
Eye,	*Naw*-chiche.	
Mouth,	Tep-pa.	
Tongue,	Ar-*ah*-ko.	
Back,	*Qua*-hee.	
Bone,	*So*-nip.	
Blood,	*Peeshe*-pah.	
Ear,	*Nah*-karke.	
Scalp,	*Pah*-pee.	
Buffalo,	*Cook*-chow.	
Ox,	Pe-mo-ro.	
Herd of horses,	Tah-*he*-yeh.	
Deer,	Ul-leek-kah.	
Turkey,	Ko-yo-*nit*-tah.	
Day,	Tah-arp.	
Summer,	*Ta*-arch.	
Winter,	*To*-han.	
Spring,	Tane-*hah*-ro.	
Night,	Too-*kah*-ra.	
Morning,	Pua-*arth*-co.	
Darkness,	Feir.	
Rain,	Er-mar.	
Snow,	Tar-kau.	
Sea,	Par-hap-hia.	
Prairie,	Pe-he-*wale*-te.	
Spring, (fountain,)	Pah-hap-pea.	
Bread,	Ta-e-*shaw*-tar.	
Melon,	Pe-*he*-na.	
Wood,	Koo-*oh*-nee.	
Forest,	Hoo-*oh*-carte.	
Bird,	*Hoo*-choo.	
Fish,	*Pa*-que.	

ENGLISH.	COMANCHE.	WITCHITA.
Snake,	Noo-*be*-er.	
Stone,	Terp.	
Lead,	*Nup*-parke.	
Pipe,	*Toh*-ish.	
Corn,	Hah-ne-*be*-teh.	
Tent,	Kah-*hah*-me.	
Wampum,	Tshe-nip.	
Kettle,	Way-he-*to*-wah.	
Boat,	Wo-we-*poke*.	
Axe,	Ho-*him*-nah.	
Spear,	Cheak.	
Knife,	Weith.	
Flint,	*Na*-da-curte.	
Shoe,	Ma-*a*-pee.	
Kettle,	*Wit*-wah.	
Town,	Kee-*nu*-kie.	
Warrior,	Too-a-vitche.	
Hot,	*Ur*-ate.	
Cold,	*Urtch*-ate.	
White,	*Too*-shop.	
Red,	A-*kop*-tee.	
Handsome,	Char-nar-bo-my.	
Live,	*Nay*-ure.	
Salt,	O-nae-*bit*-er.	
Near,	*May*-titch.	
Far off,	Ma-*nar*-kee.	
To-morrow,	Pa-*arch*-quee.	
To kill,	May-*way*-kun.	
To eat,	Tu-*kar*-roo.	
To walk,	*Her*-mumsh.	
To run,	No-*ka*-ark.	
To drink,	He-*bet*-to.	
To laugh,	Ta-*hah*-net.	
To cry,	*Tah*-kay.	
To love,	Kum-*mar*-pee.	
To trade,	Te-me-*ah*-row.	
To see,	Nah-*bo*-ne.	
To sing,	Ho-bee-er.	
To dance,	Ne-*er*-ker.	
Me,	Ne.	
You,	*Her*-che.	
He,	Sho-ku.	
They,	Pun-che.	
Very well,	O-shus-she.	
Perhaps,	Wo-har-*ke*-ne.	

APPENDIX H.—ETHNOLOGY.

REMARKS ON THE PRECEDING VOCABULARIES, BY PROFESSOR W. W. TURNER.

Of the two vocabularies here given, the *Comanche* agrees very closely with that obtained by Mr. Robert S. Neighbors, Indian agent in Texas, and published by H. S. Schoolcraft, in his History, Condition, and Prospects of the Indian Tribes, vol. II, p. 494, *et seq.;* the slight discrepancies which present themselves between the two being nearly all owing to the different manner in which the same sounds are caught and represented by different persons. The ethnological affinities of the Comanches are well known. They are the most important tribe of Indians in Texas, and constitute a portion of the great Shoshonee or Snake family, which have been led in pursuit of the buffalo far to the south of their congeners.

The vocabulary of the *Witchitas*, though less complete, is more interesting, as being the first ever published, as far as I am aware. A pretty extended examination, however, has not enabled me to discover an analogy between it and any other aboriginal tongue with which we have the means of comparison. It is true, that in Capt. Marcy's lists the words for *Osage, friend, mule, bear, prairie-dog*, are the same in this language as in the Comanche; but the entire dissimilarity of the two vocabularies in other respects, shows that the words in question must have been adopted from one language into the other, or from a common foreign source. Thus it is evident that the Comanche name for *prairie-dog* is borrowed from the Witchita, while the name for *mule* has been taken by both from the Spanish. The ethnological position of the Witchitas, then, remains still to be determined.

ALPHABETICAL INDEX.*

A.

	Page.
Abronia	297, 304
Abundance of water and grass	40
Acacia	285
Acephala	253
Acerates	290
Adair bay	117
Achillea	288
Actinella	289
Agama	234, 235
Agassizocrinus	199
Agricultural capabilities	69
Agrostis	300
Albuquerque, altitude of	112
Alecran (scorpion)	267
Alkaline character of water	8
Allionia	296, 297
Allium	298
Allium	298
Amianthium	298
Amianthus	298
Ammonites	209
Amorpha	284
Amsonia	291
Amsonia	291
Anabrus	258, 259
Anacardiaceæ	282
Analysis of gypsum spring	52
Analysis of gypsum water	91
Analysis of Marcylite	9
Analysis of subsoil	8, 23
Ananchytes	211
Anantherix	290
Anemone	280, 303
Andropogon	302
Andropogon	302
Anodonta	256
Antelopes	14, 49, 50, 62
Antilocapra	216
Antiscorbutics	36
Aphanostephus	289, 303
Aphanostephus	289
Apocynum	291

	Page.
Arachnidians	262
Araneidæ	262
Archimedipora	201
Argemone	280
Aristida	300
Aristida	300
Arkansas, geology of	179
Artemisia	287, 304
Asclepiadaceæ	290
Asclepias	290
Asclepias	290
Astarte	206
Astragalus	284
Atacamite	9, 155
Atheropogon	300
Atmospheric refraction	41
Atreus	269
Atriplex	296

B.

	Page.
Baptisia	284
Barometer broken	14
Bassaris	215
Battle between Indians	43
Batrachians	242
Bear, instinct of	57
Beaver creek	64
Beavers, habits of	33
Big Witchita	9
Birds seen	12
Bituminous coal	113
Boraginaceæ	294
Bos	216
Bottle buried	38
Bottom lands	85
Boulders	28, 186, 188
Boundary between Texas and Choctaw Nation	20
Bouteloua	300
Bouteloua	300
Bow, its use and material	98
Buffaloes	15, 25, 27, 65, 71

* Synonyms are italicised.

314　INDEX.

	Page.
Buffaloes, diminution of their numbers	105
Buffalo grass	40
Buffaloes, range of	104
Buffaloes, relation of, to Indians	103
Bufo	242
Bulimus	256
Buthus	269
Brachiopoda	201
Brachypeplus	260
Brackish water	64
Brown coal	167
Brush fence	37
Bryozoa	201

C.

Cache creek	7
Calamagrostis	300
Calligonium	296
Callirrhœ	281
Calymenia	296
Canadian	39
Canaje-Hexie	17
Canis	215
Cañons	177
Capparidaceæ	281
Carex	299
Carex	299
Caryophyllaceæ	281
Cardium	207
Cassia	285
Castilleja	292
Castor	215
Cattle-stealing	97
Centaurea	289
Centipede	272
Ceratichthys	248
Cervus	216
Chastity of prisoners, violation of	103
Chenopodiaceæ	296
Chenopodium	296
Chickasaw plum	19
Chiefhood of Indians	97
Chloris	300
Choctaw reservation, boundary of	74
Chloris	300
Chloris	300
Chondrosium	300
Chondrosium	300
Chrysopsis	289
Cirsium	289
Clematis	280
Clitoria	284
Cnemidophorus	239
Coal	180
Cnidoscolus	296

	Page.
Coal basin in west	166
Coal of Brazos	165
Cold weather	30
Coal west of Mississippi	165
Colorado, confusion of the name	4
Coluber	228
Comanches	86
Comanches, physical features of	98
Comanches, subdivisions of	94
Comanche trail	25
Comandra	294
Commelyna	297
Commelynaceæ	297
Compositæ	287
Coniferæ	297
Convolvulus	291
Convolvulus	291, 292
Convolvulaceæ	291
Copper	174
Copper, carbonate of	188
Copper ore	7, 9, 20, 155
Coronella	228
Courses and distances	139
Cretaceous fossils	181, 204
Cretaceous rocks	181
Crinoidea	199
Crotalus	217
Crotaphytus	234
Croton	295
Cross Timbers	70, 81, 84
Cruciferæ	280
Cucumis	286
Cucurbita	286
Cucurbita	286
Cucurbitaceæ	286
Cupuliferæ	297
Cyathrocrinus	199
Cyclas	254
Cyperaceæ	298
Cyperus	298
Cyperus	298

D.

Daihinia	257
Dalea	284
Dalea	284
Dasystoma	293
Deer	62, 66
Deer-bleat	50, 66
Delphinium	280
Desmodium	284
Dews	43
Didelphys	216
Dinebra	300
Diplopappus	289

INDEX.

	Page.
Distances from Fort Belknap to Santa Fé	89
Dithyræa	280, 303
Divide	12, 13, 41, 48
Dodecatheon	294
Doña Ana	113
Drift	168, 175, 185
Drift-hills	191
Drought	37

E.

	Page.
Early explorations of Red river	2
Echinacea	288
Echinodermata	210
Egletes	289
Elevations above sea	43, 56
Elk creek	20
Elymus	302
Elymus	302
Encampment, mode of	30
Engelmannia	289, 304
Engelmannia	295
Enslenia	290
Equisetaceæ	302
Equisetum	302
Eragrostis	301
Eriogonum	296
Eritrichium	294
Eritrichium	294
Eryngium	286, 303
Erythræa	291, 304
Erythræa	291
l'Etage Senonien	181
Euchroma	292
Eulima	208
Euphorbia	295
Euphorbiaceæ	295
Euploca	294, 304
Euploca	294
Eurytænia	287, 303
Eustoma	291
Eutænia	219, 220
Eutriana	300
Evolvulus	291
Evolvulus	291
Exogyra	204, 205
Explorations by Gregg	4
" " James	3
" " Long	3
" " Pike	2
" " Sparks	2

F.

	Page.
False rumors	77

	Page.
False scorpions	270
Fedia	287
Felis	215
Ferruginous sand	193
Fertility of soil	72
Festuca	301
Fimbristylis	299
Fish	24, 245
Flies, annoyance from	68, 71
Fort Arbuckle, arrival at	82
Fort Belknap	5, 11
Fort Smith, altitude of	112
Fort Smith, arrival at	5
Fossils	45
Fossil-wood	187
French explorations of Red river	2
Fresh water	37, 40

G.

	Page.
Gaillardia	288
Galeodes	270
Garter snake	219
Gasteropoda	254
Gaura	285
Gentianaceæ	291
Geraniaceæ	282
Geranium	282, 303
Gerardia	293
Geology of Arkansas	179
Geology of country	163
Glandularia	293
Globiconcha	208
Glycyrrhiza	283
Gnats, annoyance from	68
Gold	185
Gold-bearing formation	19
Gold-bearing rocks	170
Gold deposites	14
Gold diggings of Colorado	171
Grama grass	28, 43, 50
Grama	300
Gramineæ	299
Granite	14, 15, 181
Grapes	35
Grasses, native	73
Grazing, best time for	31
Gregg's expedition	4
Greyhounds, use of, in chase	25
Grossulaceæ	286
Grotto in gypsum	51
Grove of timber	21
Gryphæa	205
Guides, sagacity of	76
Gynamblosis	295
Gypsum	22, 46
Gypsum beds, extent of	91

	Page.
Gypsum deposite	172
Gypsum formation	168
Gypsum in South America	173
Gypsum water, analysis of	52, 91

H.

Head of navigation of Red river	89
Head spring of Red river	55
Hedyotis	287
Heterodon	222
Helianthus	288
Helix plebeium	28
Heliotropium	304
Helonias	298
Hemiaster	210
Hendecandra	295
Hendecandra	295
Herpetodryas	228
Hoffmanseggia	284, 303
Holaster	210
Holbrookia	235, 236
Holectypus	211
Homeward march	58
Hordeum	302
Hospitality, rights of	100
Houstonia	287
Hundredth degree of longitude	18, 19
Hymenopappus	289
Hymenopappus	289
Hymenothecium	300
Hymenoxys	289
Hypoxidaceæ	297
Hypoxis	297

I.

Ignorance of power of whites	99
Incredulity of Indians	99
Indians	76
Indians and Tartars compared	96
Indian camps	31, 33, 36
Indian forays	87
Indian horsemanship	95
Indians, mode of checking	88
Indians of Red river, general description of	93
Indian tracks	63
Indian villages	72
Indigofera	283
Inoceramus	206
Ipomaea	291
Iridaceæ	298
Iron sands	157
James, Dr., journal of	3

J.

	Page.
Janira	204
Jasper	170
Jatropha	296
Julus	274
June rise	15, 84, 91
Juniperus	297
Juniperus	297
Juniperus Virginiana	53

K.

Kaskia Indians	3
Ke-che-a-qui-ho-no	49
Keechies	93
Kickapoos	81
Kioways	37, 43, 86
Kioway creek	36
Koeleria	301
Koeleria	301
Krameria	283
Krameriaceæ	283

L.

Labiatæ	293
Latitudes	38, 56, 59, 63
Lakes of Red river	84
Laguna Colorado	111
Labradorite	157
Leguminosæ	283
Lepachys	288
Lephrosia	283
Leptocaulis	286
Leptophis	232
Lepus	218
Leuciscus	248, 249, 251
Liatris	287, 303
Liatris	287
Lies told by the Indians	18
Lightning	13
Lignite	167
Liliaceæ	298
Lime	22
Linaceæ	281
Linum	281
Lippia	293, 304
Liquor, use of, among Indians	102
Lisianthus	291
Little Witchita	5
Lizards described	233
Llano Estacado	33, 38, 39, 41, 42, 45, 49, 50, 56, 92, 114
Llano Estacado impracticable for a railroad	110

INDEX. 317

	Page.
Llano Estacado, geology of	190, 191
Loasaceæ	286
Lodges, Witchita, town of	76
Loess creek	28
Longitudes	38
Long's exploration	3
Long's peak	115
Lost member of party	39
Lutra	215
Lycosa	263
Lygodesmia	289, 290
Lygosoma	240
Lymnæa	255
Lynx	215

M.

	Page.
Magnetic needle, variation of	64
Malva	281
Malvastrum	281
Malvaceæ	281
Mammals	215
Manganese ore	157
Marcylite	9, 155
Marshallia	289
Masticophis	230
McClellan's creek	40
Medicine lodges	107
Melampodium	289
Melampodium	289
Melanthaceæ	298
Melica	301
Mentzelia	286
Mephitis	215
Meteorological observations	119
Mezquite grass	6
Mezquite wood	28, 40, 59, 66, 114
Middle Comanches	79
Military post, new one advised	87, 88
Mineralogy	155
Mirabilis	296
Mirage	41
Mollusca	204
Monarda	293
Monarda	293
Mount Scott	70
Mount Webster	21
Mountains, physical features of	65
Mulberry creek	60
Mule lost	74
Mygale	262
Myosotis	294
Myriapods	272

N.

	Page.
Navigation of Big Witchita	6

	Page.
Negroes, hostility to	101
Nemastylis	298
North Fork	24
Nyctaginaceæ	296

O.

	Page.
Oaks, dwarf	25
Obione	296
Obione	296
Œnothera	285
Oldenlandia	287
Onagraceæ	285
Ophibolus	227, 228, 229
Order for expedition	1
Orthopterous insects	257
Ostrea	205
Otter creek	14
Overcup oak	8
Oxalidaceæ	281
Oxalis	281, 282
Oxybaphus	296, 297
Oxytropis	284

P.

	Page.
Pah-hah-en-ka	79
Palæontology	199
Palafoxia	283
Panicum	299
Panicum	299
Panopœa	207
Panther	11, 50, 59, 66
Papaveraceæ	280
Paronychia	281
Paspalum	299
Pass in the mountains	70
Patent Office, letter from	60
Peak of Guadalupe	112
Pecten	204
Pentatrematites	200
Pentremites	200
Pentstemon	292, 304
Pentstemon	292
Petalostemon	283, 284
Petrified wood	42
Phalangium	298
Phalangopsis	257
Phalaris	299
Phalaris	299
Phengites	173
Phlox	294
Phrynosoma	233
Physa	255
Physalis	292
Pike's expedition	2

318 INDEX.

	Page.
Pituophys	225
Planorbis	255, 256
Plantagenaceæ	296
Plantago	296
Pleuraphis	300
Poa	301, 304
Poa	301
Polanisia	281
Polemoniaceæ	294
Polygala	282, 283
Polygalaceæ	282
Polygonaceæ	296
Polyotus	290
Polytænia	287
Pomotis	245
Porphyritic greenstone	169
Portulacaceæ	281
Poterium	285
Prairie dogs	43, 46, 59
Prairie-dog towns	46
Presents distributed	18
Preston	89
Preston, arrival at	5
Primulaceæ	294
Prisoners, release of	79
Prisoners, Mexican	79
Procyon	215
Productus	201, 202
Prunus chicasa	19
Psammophis	230
Pseudoscorpionidæ	270
Psoralea	283
Ptelea	282
Pteromys	215
Pupa muscorum	28
Pure water	64
Pyrrhopappus	289

Q.

Quapaws	93
Quercus	297
Quercus macrocarpa	8
Quicksand	7, 29

R.

Raft of Red river	84
Railroad, best route for	110, 112
Railroad, general considerations respecting	109
Rain	13, 14, 15, 65
Rains, times of occurrence	42
Rana	243
Ranunculaceæ	280
Rattlesnake	217

	Page.
Red clay formation	168
Red river	13
Red river, early explorations of	2
Red river, its physical characters	83
Red river, position of sources	84
Reptiles	217
Reptiles collected	61
Religious belief	107
Retepora	201
Rhus	282
Rhynchosia	283
Ribes	286
Riddellia	288
Riddellia	288
Rio Raijo of Humboldt	4
Rise of river	44
River terraces	90
Rock bed of river	54
Rock salt	91
Rosaceæ	285
Rubiaceæ	287
Rudbeckia	288
Rudbeckia	288
Rush creek	80

S.

Sabbatia	291
Safe return	82
Sagacity of Indians	32
San Diego	116
Sand-hills	16, 90
Sandy creek	39
Salt Fork	21
Salt, incrustation of	35
Salt springs	181
Salt plains not at head of Red river	42
Sanguisorba	285, 303
Santalaceæ	294
Sapindaceæ	282
Sapindus	282
Sceleporus	236
Scenery, magnificent	55
Schrankia	285
Schrankia angustata	44
Scilla	298
Scincus	241
Sciurus	215
Scolopendra	272
Scoria	169
Scorpio	267
Scorpions	267
Scorpionidæ	267
Scotophis	226
Scrophulariaceæ	292
Scutellaria	293
Scurvy	36, 44, 68

INDEX. 319

	Page.
Scyphia	168
Sections, geological	182
Selenite	187
Septaria	158
Serpents	217
Sesleria	301
Shells	253
Shepard, Prof. C. U., letter from	155
Shortest route to Pacific	115
Sicydium	286
Sicydium	286
Sida	281
Sidell's route	115
Sierra Waco	113
Signs, use of	103
Silene	281
Sisyrinchium	298
Smoke of Indians	62
Snows, little obstruction from	114
Soils, analysis of	158
Solanaceæ	292
Solanum	292
Solidago	287
Sparks's expedition	2
Spermophilus	216
Spiders	262
Spirifer	203
Sporobolus	300
Staked Plain	56
Staked Plain, etymology of	92
Stevia	288
Stillingia	295
Stillingia	295
Streptanthus	280
Subsoils	176
Subsoil, analysis of	8
Succinea	256
Succinea elongata	28
Sulphate of lime	45
Sulphur river	113
Sulphuret of lead	80
Superstitions	107
Suydam creek	25
Sweetwater creek	29
Swiftness of wild animals	62, 63

T.

	Page.
Talinum	281
Talk with the Indians	17
Tamias	215
Tarantula	262
Tarantulidæ	265
Tartars and Indians compared	96
Telegonus	267
Temperature of water	8, 185
Terebratula	202, 203, 207

	Page.
Terraces of river	35
Tertiary coal	167
Teucrium	293
Teucrium	293
Thelyphonus	265
Thesium	294
Thirst	53, 54
Tiliqua	241
Tillable land on Red river	86
Timber	65, 70, 73
Timber, large size of	40
Timber, varieties of	8, 12
Titanian sands	157
Tobacco, use of	102
To-se-quash	79
Tradescantia	297
Traffic of Indians, illegal	105
Tragia	295
Tragia	295
Transportation of stores, route for	89
Trigonia	206
Trinity river	113
Tripsacum	302
Triticum	302
Tylostoma	208

U.

	Page.
Umbelliferæ	286
Unio	253, 255, 256
Uniola	301, 304
Uniola	301
Ursus	215

V.

	Page.
Valerianaceæ	287
Valley, fertile	70
Verbena	293
Verbenaceæ	293
Vermillion, use of	99
Vesicaria	280
Vicia	283
Vinaigron (*Thelyphonus*)	265
Vitaceæ	282
Vitis	282
Volcanic rocks	169
Vulpes	215

W.

	Page.
Wacos	77, 78, 93
War-club	98
War parties	97

	Page.
War parties, how distinguished	25
Warner's pass	116
Water basin	59
Water of Red river, analysis of	176
Water, sudden rise of	12
Wild cat	61, 101
Wild-horse creek	80
Winds, prevailing	30
Winters of Red river	86
Witchitas	17, 77, 93
Witchita mountains	10, 15, 62, 64
Witchita mountains, agricultural capabilities of country about	73
Witchitas, extent of their country	69
Witchita mountains, structure of	186
Women, condition of	102

X.

	Page.
Xanthisma	303

Z.

	Page.
Zanthoxylaceæ	282
Zapania	293
Zinnia	288
Zoology	215

GRANITE BOULDERS.

75 feet in circumference near the western extremity of the Witchita mountains

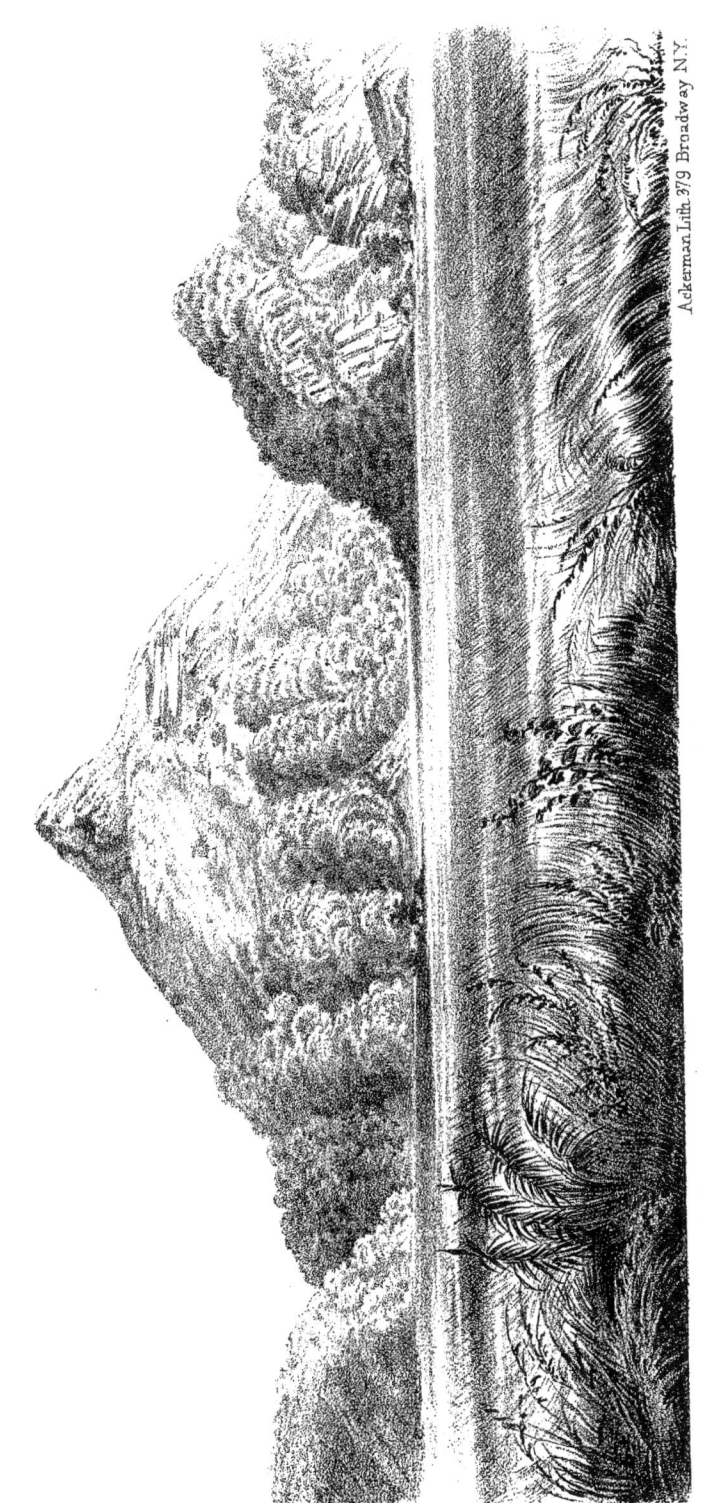

MOUNT WEBSTER.

Ackerman Lith. 379 Broadway N.Y.

ENCAMPMENT ON 6th JUNE.

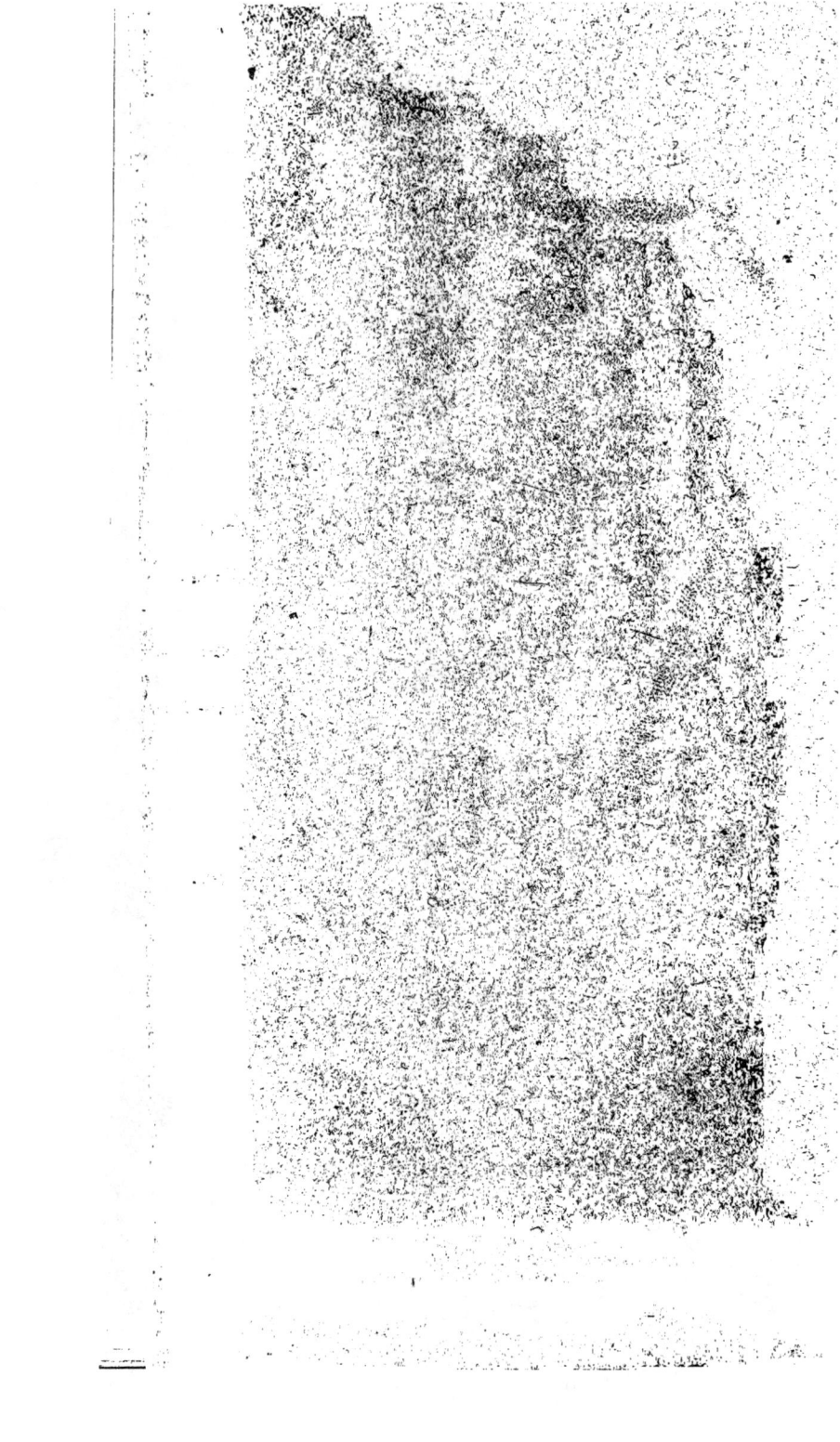

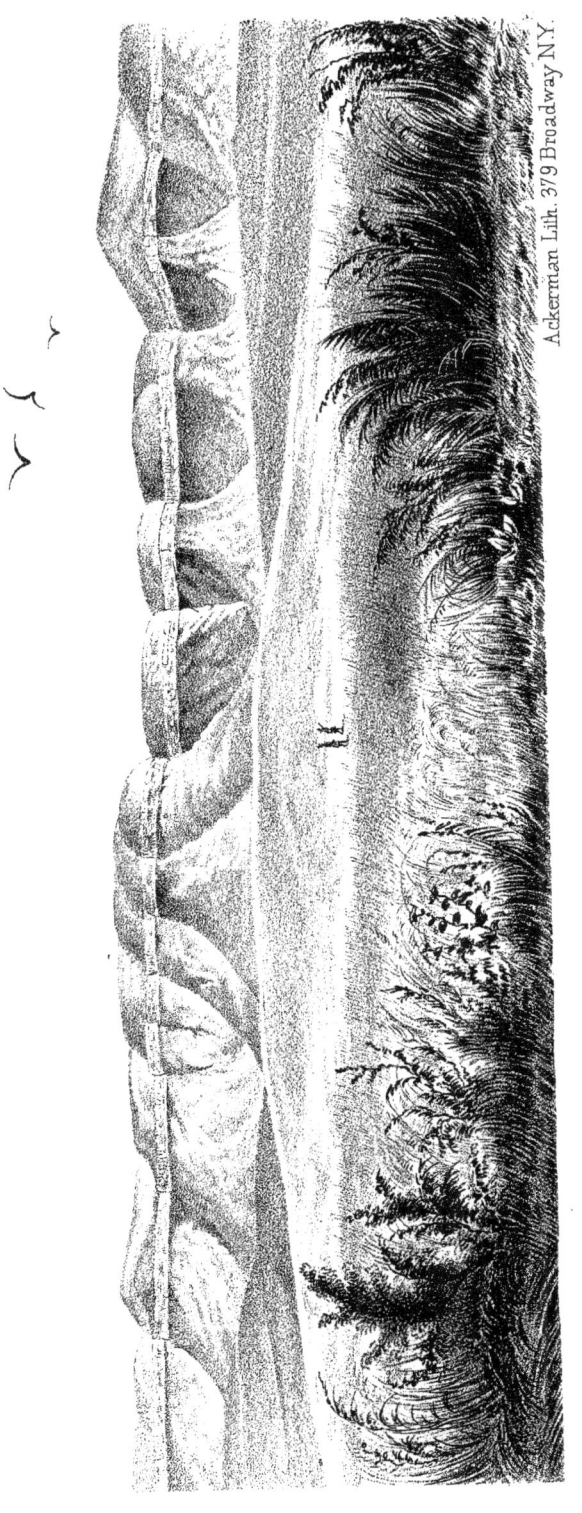
GYPSUM BLUFFS ON NORTH BRANCH RED RIVER.

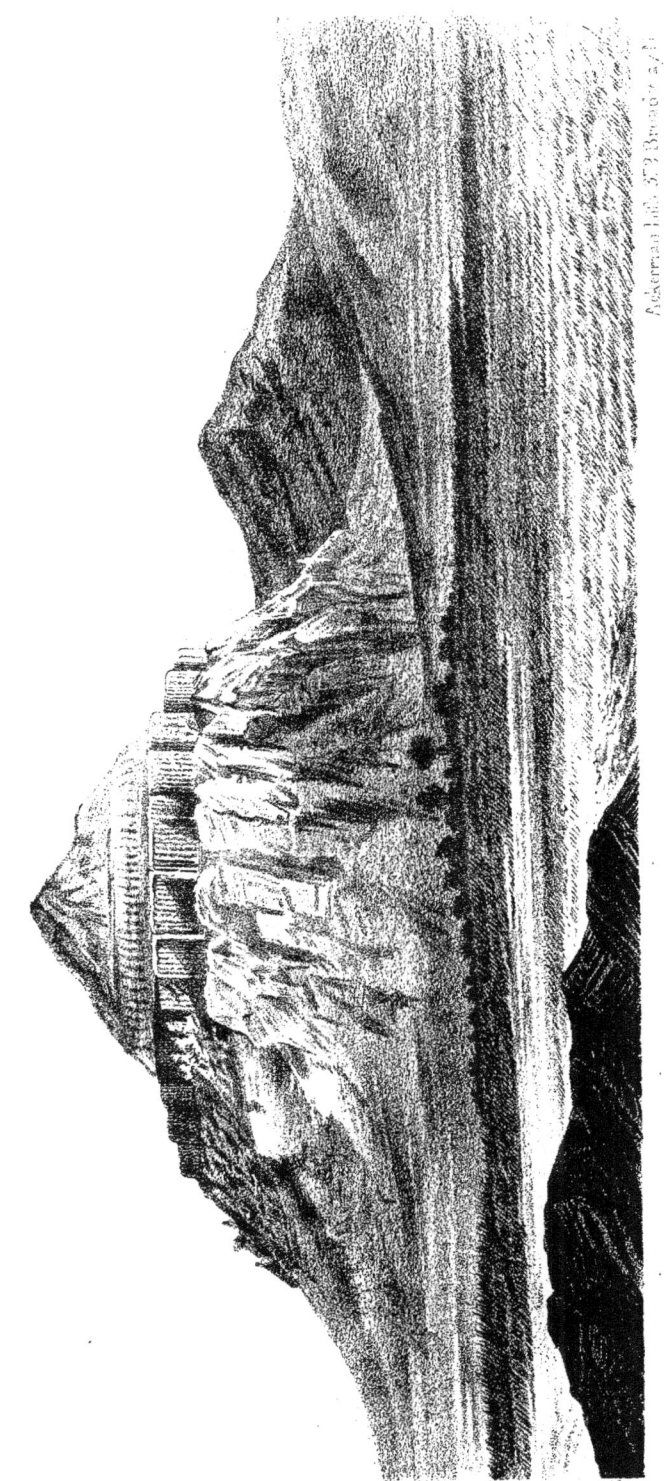

VIEW NEAR GYPSUM BLUFFS ON RED RIVER.

BORDER OF EL-LLANO ESTACADO.

VIEW NEAR THE HEAD OF THE KE-CHE-AH-QUE-HO-NO

VIEW NEAR HEAD OF RED-RIVER.

HEAD OF KE-CHE-AH-QUE-HO-NO OR THE MAIN BRANCH OF RED-RIVER

CPSIA information can be obtained
at www.ICGtesting.com
Printed in the USA
LVHW061013100319
610086LV00059B/687/P